Art in the Archaeological Imagination

Art in the Archaeological Imagination

edited by
Dragoș Gheorghiu

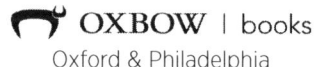
OXBOW | books
Oxford & Philadelphia

Published in the United Kingdom in 2020 by
OXBOW BOOKS
The Old Music Hall, 106–108 Cowley Road, Oxford OX4 1JE

and in the United States by
OXBOW BOOKS
1950 Lawrence Road, Havertown, PA 19083

Paperback Edition: ISBN 978-1-78925-352-8
Digital Edition: ISBN 978-1-78925-353-5 (ePub)

A CIP record for this book is available from the British Library

Library of Congress Control Number: 2020931800

Printed in the United Kingdom by CMP (UK) Limited

Typeset by Versatile PreMedia Services (P) Ltd

For a complete list of Oxbow titles, please contact:

UNITED KINGDOM
Oxbow Books
Telephone (01865) 241249
Email: oxbow@oxbowbooks.com
www.oxbowbooks.com

UNITED STATES OF AMERICA
Oxbow Books
Telephone (610) 853-9131, Fax (610) 853-9146
Email: queries@casemateacademic.com
www.casemateacademic.com/oxbow

Oxbow Books is part of the Casemate Group

Front cover: Land art revealing the walls of a Chalcolithic castro on Monte Velho, Portugal (2010–2013), by Dragoş Gheorghiu. On the right the archaeological excavation of the walls. (Photo by D. Gheorghiu)

Back cover: A bifacial pointed stone artefact, Middle Stone Age, from Blombos Cave, South Africa (photo: Magnus Mathisen Haaland)

In memoriam Dr. George Gheorghiu (1919–1969)

Contents

Contributors

TIMOTHY DARVILL
Professor, Department of Archaeology
and Anthropology, Faculty of Science
and Technology, Bournemouth University,
Fern Barrow, Poole, Dorset BH12 5BB. UK

ROBERTA ROBIN DODS
Associate Professor Emeritus, University of
British Columbia Vancouver, and Irving K.
Barber School of Arts and Science, University of
British Columbia Okanagan, Canada

DRAGOS GHEORGHIU
Professor, Doctoral School, National University
of Arts, Bucharest, Romania, and Earth and
Memory Institute (ITM), Quaternary and
Prehistory Group of the Geosciences Centre
(u. ID73–FCT), Portugal

TORILL CHRISTINE LINDSTRØM
Professor, Department of Psychological Science,
Faculty of Psychology, and SapienCE,
Centre for Early Sapiens Behaviour, CoE,
Faculty of Humanities, University of Bergen,
Norway

GEORGE NASH
Associate Professor, Geosciences Centre of
Coimbra University (u. ID73-FCT),
ITM (Earth and Memory Institute),
Polytechnic Institute of Tomar (IPT), Portugal
E-mail: George.nash@IPT.pt

ELIZABETH PORAJ-WILCZYNSKA
E-mail: lizpw@yahoo.co.uk
Website: http://lizpw.co.uk/

IEGOR REZNIKOFF
Emeritus Professor, Université de Paris Ouest
Nanterre, France
E-mail: dominiqueleconte@yahoo.fr

JACQUI WOOD
Saveock Water Archaeology, Truro, UK

EZRA ZUBROW
Professor, Department of Anthropology,
University at Buffalo and University of Toronto,
Canada

Introduction

Dragoş Gheorghiu

Walking away from the neatly bounded, regular, categorically defined ontologies of positivism and western science as practiced over the past 300 years, past the dung-heap of post-modernist self-referential hyper-critical pluralism, we find ourselves in new fields where thinking is more fluid, more porous, and more emotionally stimulated. Gone are the constraints of Cartesian Dualisms, binary oppositions, and simplistic back-projections of modern life onto other people's existence. (Darvill & Poraj-Wilczynska, *this volume*)

Art and humanities – a fertile relationship

During the last decades of the 20th century the barriers between art and science and especially the humanities began to be blurred (Gheorghiu & Ştefan 2013; Malina 2016). Starting in the 70s, contemporary artists have adopted an 'anthropological' view, drawing inspiration from the fields of anthropology (see Foster 1994), and archaeology (see examples in Renfrew 2003). The 'artist as anthropologist' (Kosuth 1993) signifies an immersion of the artist and his social engagement. This period coincides with the crisis of representation in anthropology and its opening towards new ways of seeing. The tendency towards visual anthropology became visible from the 1980s, along with the crisis of representation 'and the (re) establishment of visual, sensory and applied anthropologies' (Pink 2006, 12 ff).

The use of visual media in anthropology (Banks & Morphy 1997) has brought it closer to contemporary artistic techniques. In the newly created relationship 'productive dialogues between the domains of contemporary anthropology and art' have been generated (Schneider & Wright 2006, 1).

Art attracted anthropologists because it is 'irrational, mysterious, [and] numinous' (Hiller 1991, 2).

For anthropologists to engage with art practices means embracing new ways of seeing and new ways of working with visual materials. This implies taking contemporary art seriously on a practical level and being receptive to its processes of producing works and representing other realities. (Schneider & Wright 2006, 25)

The next step in contemporary anthropology was engaging the senses (Pink 2006).

What happened in archaeology? How did the change and tendency to art occur? In archaeology too there was a 'crisis of representation' as in anthropology; it prompted the decline of Processualism and represented a change similar to that which imposed the anthropology of the visual and of the senses, and opened the archaeological discipline to art. The post-processual tendencies that emerged in the last decades were characterised by a predisposition to art: these new ways of seeing the world discussed symbolism, the phenomenology of the landscape, or the materiality of the world.

This propensity to art was a slow progression, starting from approaches to symbolism (Hodder 1982), to the experientiality of the world (Shanks 1992) and metaphors (Tilley 1999), to the phenomenology of landscape (Tilley 1994), to theatre archaeology (Pearson & Shanks 2001), senses (Gheorghiu 2009a; Fahlander & Kjellström 2010; Skeates 2010), and the materiality of the world (Tilley 2004).

All these new visions that led to the discovery of an 'archaeological art' (Shülke 2000; Tilley *et al.* 2000; Hamilakis *et al.* 2001; Cochrane & Russel 2007; Gheorghiu 2009b; 2009c; 2012a; 2012b; Russel 2011; Russel & Cochrane 2014; Chittock & Valdez-Tullett 2016; Bailey 2018; Gheorghiu & Barth 2019) issued from the trends of contemporary art and from the ones that study the art of the Past (Jones & Cochrane 2018), together considerably changed our image of the Past, and influenced the archaeological imagination (Gheorghiu & Bouissac 2015). All this orientation of archaeology to art shows that, as an analogical way of thinking, art has existed in the mind of archaeologists. To this, one can add the Gestalt principles and the psychological response to art which is emotion (Matravers 2001), the basis on which imagination functions.

Consequently, the purpose of this book is to present the archaeological research functioning as a sort of artistic creation, proposing new perspectives on the archaeological imagination. It offers an exploration of the creative processes, the possibility of finding inspiration in experientiality, and the approach to the act of creation as a subject for archaeological research. When analysing the art of the Past, or when using art methods to approach the Past, we are facing an act of creation where imagination, emotion, and creativity combine under the form of an experiential instrument of investigation.

The book offers a vision of archaeological research, a means to understand the complexity of the human nature and, consequently, to approach the human thinking structured on similarity and symbolism, being able to detect cultural and psychological subjects ignored until today and, at the same time, to offer a series of visions of art, seen from the perspective of archaeology.

The common concept that links all the chapters of this book is the archaeological imagination, a creative process analogous to the artistic imagination. As Michael Shanks (2012, 149) observed in his seminal book on this subject: 'at the heart of the archaeological imagination is *creative* practice that cuts across science and humanities, the past and the present'.

The present volume follows David Bohm's (2007, 105) '*art form* of science', namely the presentation of archaeological thinking as a form of art, revealing the poetics of the archaeological imagination. It shows that, in their work, archaeologists, without being inspired by contemporary artists, use creative methods, and their analysis of the art of the Past goes beyond the material culture of the art objects, into the realm of the mental processes of creation.

The book is structured into three parts that complement each other.

Imagining the art of the past as a magic and emotive experience

The Past is preserved in the form of material objects or oral traditions to literate forms, which are brought back to life by the archaeological imagination that re-embodies them, tending to reach the 'tonal quality and rhythmic fluidity' of the ancient worldviews (Dods, this volume).

How can we approach the objects we call art from the archaeological record? This is a double subjective experience that attempts by reverse engineering 'to reach an understanding of the maker's mind' and thus to imagine 'the subjective experience of that' human 'mind of the Past' (Dods, this volume). Also subjective is the experience of aesthetic forms (Lindstrøm, this volume), which generates emotion and is revelation; it can reveal the skill of the artist but also some characteristics of Nature, such as the laws of efficiency (Gheorghiu, this volume).

In archaeological imagination the artistic research tries to reach the psychological factors proper to the makers of these objects. Objects that are characterised by aesthetic attributes as balance and proportion, or symmetry and Gestalt, imply sensory interactionism with the receptor, as first stated by Plato (*Phaedrus*) (Lindstrøm, this volume).

Experimentation and experience as forms of art

One of the current methods of approaching the Past is to experiment with its material culture. In this case, the action of the experimentalist becomes a work of art of the type of the ancient *technē*, a concept that simultaneously expresses the technical and aesthetic mastery of the resulting product.

Also *technē* could be found in the efficiency of the design of some patterns or objects (Gheorghiu, this volume). Experimenting the aesthetics of the Past implies an interactionism of the individual with the work of art, in some cases requiring a sensory intervention, as in the production of sounds in front of the rock paintings positioned in areas of maximum resonance. Such eco-relationship that is created between images and performer 'is an unforgettable experience' (Reznikoff, this volume).

In principle, an experimental archaeology approach is a mimetic, faithful copy of an object, using a reverse engineering process (Wood, this volume). The experimenter has in mind the model he/she tries to reproduce as accurately as possible, in all its technological and material details. At the same time, a work of experimental archaeology is also an experiential one, both of objects and of their contexts.

The phenomenological experience of the landscape and environments can be visualised through artistic representations 'based upon sensory and immersive experiences' and through this type of experience the humanity of the Past can be linked to present-day imaginations (Darvill & Poraj-Wilczynska, this volume). The bodily memory gained through sensory experiences can later reproduce in the mind of the experimenter visual patterns or tactile sensations. It is to this type of memory and sensorial imaginary that the mind of the visitor appeals when immersed in the art of the virtual worlds, where the experimentation of the materiality of the spaces is a disembodied process, in order to augment the bodily experience (Gheorghiu, this volume).

The exploration of the act of creation

When exploring the act of creation, one cannot 'ignore the similarities between archaeology and the physical act of painting', since the artistic research offers a new perspective of how archaeological imagination can be seen as a new form of (history of) art. The artist acts as a 'history-maker', creating a 'multifaceted image', therefore an artwork is similar to a Harris Matrix, from the point of view of its making (Nash, this volume). The definition of the artist, as well as that of the art, avoids precise descriptions; in addition it creates a series of diverging questions. However, modern society has managed to divide and classify the creative act into countless categories, just as it did with Nature's creation.

How can archaeology define the creative act, or how can it imagine the prehistoric artists? One approach might be provided by the use of heuristic substitution methods (Zubrow, this volume).

The book ends with a set of questions valid both for the definition of prehistoric and modern artists and which highlights the complexity of the creative act.

References

Bailey, D. 2018. *Breaking the Surface. An Art/Archaeology of Prehistoric Architecture*. Oxford: Oxford University Press.

Banks, M. & Morphy, H. (eds) 1997. *Rethinking Visual Anthropology*. Newhaven and London: Yale University Press.

Bohm, D. 2007. *On Creativity*. London and New York: Routledge.

Chittock, H. & Valdez-Tullett, J. (eds) 2016. *Archaeology with Art*. Oxford: Archaeopress.

Cochrane, A. & Russell, I. 2007. Visualising archaeologies: A manifesto. *Cambridge Archaeological Journal* 17, 3–19.

Fahlander, F. & Kjellström, A. (eds) 2010. *Making Sense of Things. Archaeologies of Sensory Perception*. Stockholm: Stockholm University Press.

Foster, H. 1994. *The Return of the Real: Art and theory at the end of the century*. Cambridge MA: MIT Press.

Gheorghiu, D. 2009a. *Artchaeology. A Sensorial Approach to the Past*. Bucharest: Unarte.

Gheorghiu, D. 2009b. De l'objet à l'espace: une expérience art-chéologique de la préhistoire. *Etudes Balkaniques* (Cahiers Pierre Belon 15), 211–224.

Gheorghiu, D. 2009c. A study of art-chaeology. *Archeologia Africana - Saggi occasionali 2005–2009*, 11–15, 45–50.

Gheorghiu, D. 2012a. eARTh Vision (Art-chaeology and digital mapping), *World Art* 2(2), 211–217.

Gheorghiu, D. 2012b. Metaphors and allegories as Augmented Reality. The use of art to evoke material and immaterial subjects. In I-M. Back Danielsson, F. Fahlander & Y. Sjöstrand (eds) *Encountering Imagery. Materialities, Perceptions, Relations.* Stockholm: Stockholm Studies in Archaeology 57, 177–185.

Gheorghiu, D. & Barth, T. (eds) 2019. *Artistic Practices and Archaeological Research.* Oxford: Archaeopress.

Gheorghiu, D. & Bouissac, P. (eds) 2015. *How do we Imagine the Past? On Metaphorical Thought, Experientiality and Imagination in Archaeology.* Newcastle-upon-Tyne: Cambridge Scholars Publishing.

Gheorghiu, D. & Ştefan, L. 2013. In Between: Experiencing liminality. *Leonardo Electronic Almanac* 19(1), 44–61.

Hamilakis, Y., Pluciennik, M. & Tarlow, S. 2001. Academic Performances, Artistic Presentations. *Assemblage* 6. [https://archaeologydataservice.ac.uk/archives/view/assemblage/html/6/art_web.html]

Hiller, S. 1991. Editor's introduction. In S. Hiller (ed.) *The Myth of Primitivism.* London: Routledge.

Hodder, I. 1982. *Symbolic and Structural Archaeology.* Cambridge: Cambridge University Press.

Jones, A.M. & Cochrane, A. 2018. *The Archaeology of Art. Materials, Practices, Affects.* London and New York: Routledge.

Kosuth, J. 1993. *Art After Philosophy and After: Selected writings.* Cambridge MA: MIT Press.

Malina, F.R. 2016. Art-science: An annotated bibliography. *Art Journal* 75(3), 64–69.

Matravers, D. 2001. *Art and Emotion.* Oxford: Clarendon Press.

Pearson, M. & Shanks, M. 2001. *Theatre/Archaeology.* London: Routledge.

Pink, S. 2006. *The Future of Visual Anthropology. Engaging the Senses.* London and New York: Routledge.

Renfrew, C. 2003. *Figuring it Out. What are We? Where do We Come From? The Parallel Visions of Artists and Archaeologists.* London: Thames and Hudson.

Russel, I. 2011. Art and archaeology. A modern allegory. *Archaeological Dialogues* 18(2), 172–176.

Russel, I. & Cochrane, A. 2014. *Art and Archaeology. Collaborations, Conversations, Criticisms.* New York: Springer.

Schneider, A. & Wright, C. (eds) 2006. *Contemporary Art and Anthropology.* Oxford and New York: Berg.

Shanks, M. 1992. *Experiencing the Past: On the Character of Archaeology.* London: Routledge.

Shanks, M. 2012. *The Archaeological Imagination.* Walnut Creek CA: Left Coast Press.

Shülke, A. 2000. Archaeology and art. In C. Holtorf & H. Karlsson (eds) *Philosophy and Archaeological Practice. Perspectives from the 21st century.* Göteborg: Bricoleur Press, 171–273.

Skeates, R. 2010. *An Archaeology of the Senses. Prehistoric Malta.* Oxford: Oxford University Press.

Tilley, C. 1994. *A Phenomenology of Landscape: Places, path and monuments (explorations in archaeology).* Oxford: Berg.

Tilley, C. 1999. *Metaphor and Material Culture.* Oxford and Malden: Blackwell.

Tilley, C. 2004. *The Materiality of Stone.* Oxford: Berg.

Tilley, C., Hamilton, S. & Bender, B. 2000. Art and the re-presentation of the past. *Journal of the Royal Anthropological Institute* 6, 35–62.

Chapter 1

Reveries and representations of the magic of being

Roberta Robin Dods

> *... past is prologue* (Shakespeare: *The Tempest*)

Much of our contemplations of the pre-literate past come to us through artefacts, objects, defined as utilitarian, artistic, symbolic/'religious', etc, and contextualised on their essential components, production patterns and environmental placement in time and space, essentially through *hardware* classification (typology/taxonomy). Such ordering has been (and continues to be) used in the building of frameworks for synchronic and diachronic cultural elucidations. As a metaphor *hardware* suggests that we are defining things in either/or statements while on consideration of our actual data we have superseded the binary digital approach transcending to a model/metaphor found in the **qbit** (aka qubit) (Schumacher 1995) that allows us to contemplate beyond either/or to both/and statements. Material culture is what I consider to be 'texts', the code and, in a sense, Schumacher compressions – on the surface seemingly relatively simple presentations of very complex intersections of materiality created by the dynamic relationship of our physical beingness with our diverse and challenging environments. Consider as a metaphorical **qbit** Marshack's interpretation of 'lunar phrasing' on a marked piece of bone (Rosenberg 2007–2008). The deep complexity rests in the feedback loop(s) created by the *in there/out there* interactions of the conscious mind of the maker with an exceedingly dynamic world and *vice versa*. This is also so for Marshack's process of coming to understanding deep complexity and thus illustrating a mind in the process of describing/scribing something significant in the life (Rosenberg 2007–2008). Renfrew had already given this direction in *Material Engagement Theory* (2004) where he proposed a new paradigm for understanding material culture, noting that material culture, in reality, substantiates its own meaning.

Some of these 'texts' are site specific, non-portable, such as the great cave paintings of the Upper Palaeolithic. But then again there are portable objects, most famously the various Venus figurines, setting aside tools in this instance, that in many occurrences charge us to seriously consider our engagement in the symbolic representation of the world perceived. Like whispered poetry things can be exceedingly ephemeral if they are not associated with specific site locations. Objects also have the potential of being lost in the archaeological taphonomy of retrieval, lab analysis and storage (Bello *et al.* 2013). Here 'Including the archaeologist in the taphonomic interval gives symmetry to the investigation. This symmetry brings the culture(s) of the past into articulation with the culture of archaeology through the agency of humans' (Dods 2017). This is important to our understanding of process as interpretation that arises from the archaeologist's own selectively (overt or covert/conscious or unconscious) created cognition in the culture/environment of archaeology.

Material culture is all 'text' for pre-literate peoples. It is supported in community by the oratory (or performance piece set to music, a living 'opera', in the Latin sense) of those (elders?) who hold in memory the stories of the past and the lessons/patterns of reification as found, for example, with fire domestication, knapping of stone tools or the fashioning of an object with emotive and/or symbolic meaning. Survival rests in the transgenerational endurance of knowing ***the way*** things are to be done (Why? Asks the child. 'Because' answers the mother). Much like the Japanese concept of –***do*** (Dods 2004), [and 生き甲斐 (*ikigai*)], where the essence of the idea is the reason for being – finding value/purpose in the doing. Memory, myth and legend sustains but as cultures fade and disappear, they eventually only remain with us through reified objects we call artefacts and the conversion of oral traditions to literate forms. These are ideas to tease us through the generations. They tickle our imagination and, sometimes, pain our funny bone. Deep meaning is elusive and our enculturation may not provide us with the necessary entrée.

Some illuminative examples, European in origin but not standing apart from wider representation through other cultures, are early texts of ancient thoughts. *Beowulf*, with 3,182 alliterative lines in Old English, brings to us one part of an oral tradition past. So too do the minstrel poems of Nordic mythology, today best represented in the Icelandic *Codex Regius* (*Poetic Edda*) (Hollander 1962). With the Nordic materials there is a long complicated literary history but the minstrel remains in the poetic metre that gives us voice and the more so when recited (read) aloud. The alliteration and the cadence of the orator resonate still. How do we remember? We make it in the spirit, music, give it rhythm and in symbolism of life and death make it magic. The *Völundarkviða* (swan maidens) are a force in the *Edda* and, as with many cultures (Miller 1987), they are shapeshifters, transformers, transcendental. Our knowing of their imagery comes from deep time: 'Grave 8' of the Vedbaek Burial (Denmark) where 'the child who had his life on earth was then offered a blessing in his afterlife. The swan's wing would carry him into a better after world'.[1] This burial of a child on a swan's wing resonates in the recent poem of Robin Robertson (2011), *Grave Goods*:

> ... A seated woman with a baby
> In her lap, dusted in red ochre, next to a man
> Wearing a crown of antlers. Between the two,
> And dead like them, a young child laid down
> Into the wing of a swan.

And with us still in Tchaikovsky's *Swan Lake* is the life and death, good and evil theme. Here is a trope of longstanding and our dream of flying or transcending our natural selves. Our dreams turn to communing with and within the transcendental, to reaching for the clouds, seeking the rain, catching the fire of the day and of the storm, seeing time emerge with the moon and the movement of the stars (Jung 1933; Campbell *et al.* 1988).

As easily we could turn to the ancient *Mesopotamian Code of Hammurabi* (1754 BCE) and its 282 laws in the Akkadian language written in Sumerian invented cuneiform on recyclable sun-dried clay tablets. Here the structure of social interaction and the means to live in society becomes collected and arranged. The spell cast by the magic of the voice is 'spelled' into a stabilised form – a very linear casting (solidification) of a more fluid and musical, but nonetheless compelling, past. An earlier version, seen in *The Hazor Law Code* fragments (Israel), perhaps originating in a pre-literate time, is a socio-cultural document. Here are the words of how we are to live together and they eventually come through the Judeo-Christian tradition as a set of 'thou shalt not'. Literacy resulted in a qualitative and quantitative shift in 'information' but, theoretically, a diminishment of its tonal quality. Voicing reinstates gravitas through the resonance of the tonal quality of the message.[2] Thus, experience is re-embodied. Pre-literate ancient contemplations in material form have been addressed by us as 'art' and not infrequently considered through various analytical references to our lives in the natural and/or transcendental worlds we assume we have access to and inhabit in all current diversity. But have we thought of sounds of description and awe they experienced for the objects we now retrieve? Let them sing in our hearts and not dismiss our shared humanity. However as archaeologists, not infrequently we approach early material-based memories through the lens of the culturally constructed and technologically expanded categories that we consider important to our understanding of analysis and interpretation. As such, theory and methodology range from material analysis to the ascetic musings of art as we understand and define it and we may miss the tonal quality and rhythmic fluidity of their lives living again in our imaginations.

Foreground to background

For our ancestors sometime in the very distant past there was a significant quantitative and qualitative shift in the capacity of the brain. In the context of Caleb Scharf's article on the observable universe and his distinction between atoms that have cycled through the sensate state and those that have not, I want to reflect on intent as we 'experience that slippery property known as consciousness' (Scharf 2017). The enactivist view can be a practical emergent approach to our understanding of how cognition arises. Enactivism reconciles the outcome of the organism's (our) dynamic interface with environment(s). This is a reflection on the interactional nature found in the whispers from the deep past and yet are still core to the seeking of self and the nature of being sentient through time and space. '[C]ognition is not the representation of a pre-given world by a pre-given mind but is rather the enactment of a world and a mind on the basis of a history of the variety of actions that a being in the world performs' (Varela *et al.* 1992, 9). So how do we come to understand our emergence as beings with a cognitive 'life'? We can look at how we came to our current understanding and then turn, in this instance, to consider an application of our learnings to the topic we call 'art'. We must be willing to consider a more expansive enactment of the human imagination, heart, spirit than we sometimes give credence since it is possible lives can be lived as poetry, finding even in the commonplace a deep sense of gratifying engagement of self and community (Fig. 1.1).

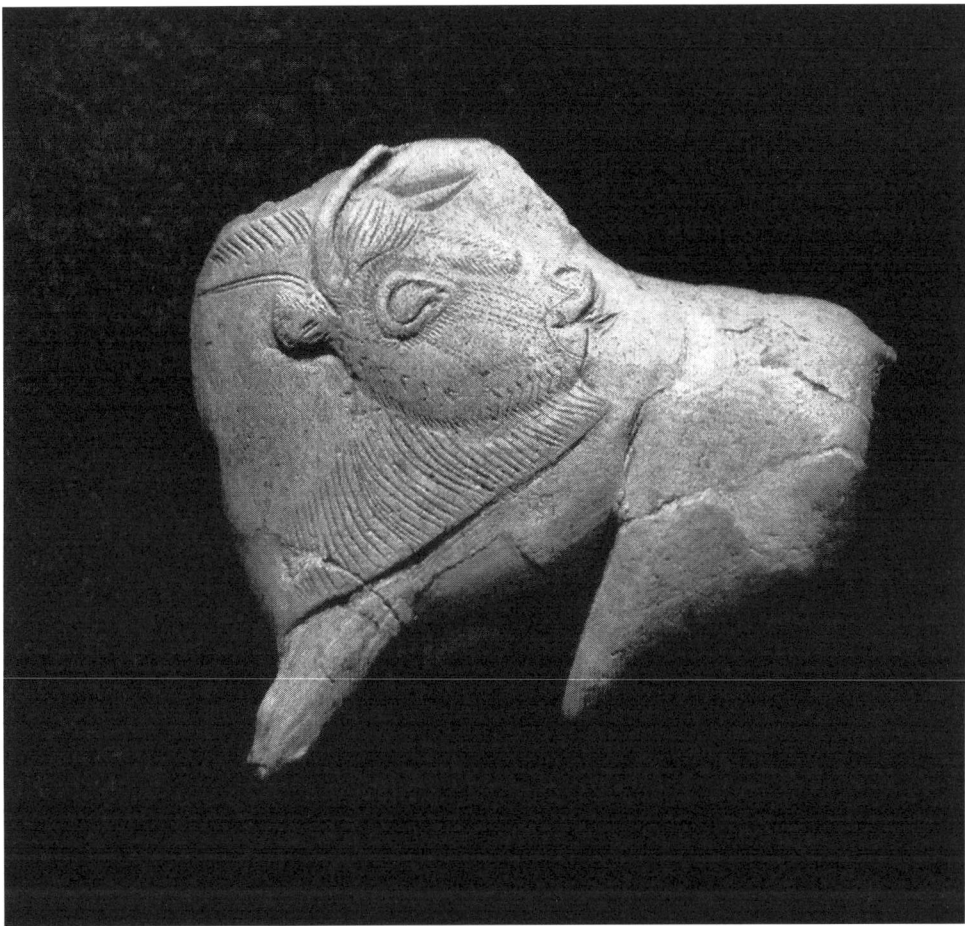

Figure 1.1: Reindeer antler fragment carved as a Bison. The animal is frequently described as nibbling an insect bite. However, from the author's direct experience of Bison this is actually a Bison licking its back after using a 'wallow'. Such locations provide minerals and mineral salts required for the Bison's diet. Note the projection of the animal's tongue. (National Museum of Prehistory, France. Les Eyzies-de-Tayac-Sireuil)

During the Age of Enlightenment, we began to see ourselves in the worlds of the past, albeit *through a glass darkly*, yet alluding to biblical interpretation. Increasingly, such interpretations were being challenged. Following from the 18th century significant challenges to our understanding of the past occurred. Momentous was the expansion of 'time' from an emerging understanding of geology based on stratigraphy (so much for Ussher's 22 October 4004 BCE), the discovery of ancient 'human' remains (specifically Neanderthals) and tools representative of diverse technologies utilising specific raw materials. Markedly influential was the work of the co-discoverers of natural selection, Darwin and Wallace, thus the theory of evolution. Certainly, our knowledge base increased while the scientific method was restructured with more investigated vigour. However, in retrospect, some interpretations were less than helpful. For example, deliberate on Van Reybrouck's *Imaging and Imagining the Neanderthal: the role of technical drawings in archaeology* (1998). He provides an effective example of interpretation bias resulting from social conventions of the 19th century. He discovered that the use of the *camera lucida* in conjunction with specimen orientation and shading techniques produced a drawing of a recently found Neanderthal skull with an aspect in compliance with the existing 19th-century theories about the past. Content analysis revealed that the supposedly objective, systematic and qualitative presentation of the Neanderthal in what was considered a 'scientific' mode actually only reinforced of the concept of 'primitive'. Interestingly, Errington (1998) gives us a timeline for the development of the concept of 'primitive' as applied to art, and art as part of the development of the capitalist market for these 'products' as 'others' and their material objects became curiosities for the peoples of the colonising nations. Thereby the Age of Enlightenment contributed 18th-century concepts of progress to 19th-century interpretations. This then became disciplinary tenets, in some instances through to the mid-20th century. Accompanying this was the challenge of the changes in technology, driving us to and reinforcing our ideas of modernism and a way of life we could contrast to physical depictions from palaeontological finds and the art and artefacts of the 'other'. Rather, we need to consider that all people live in their own 'modern times', their own assemblage of advantages and constraints contingent on their place and time. Nonetheless, the problem of presentation through supposedly a value-free medium continues today. The distinction that emerges is the trope embodied in concepts of 'primitive' as used by those in positions of relative power.

Yet there has remained a somewhat accepted cliché that the camera never lies. Of course, it can dissemble two ways: through the process of commission and the process of omission. Framing, the creation of a cognitive boundary, includes or excludes specifics in the creation of a desired result, a result that may become embedded as a meme. The upshot is that the subject becomes 'us', our analysis, representation and frequently nothing more than 'Kilroy was here', 'look where I am', or evidential documentation such as with Malinowski's self-portraits of fieldwork where the pictures according to Bell and Smith '... provide[s] in many senses harder evidence of his visit to the Trobriands than the stamp on his passport' (Ball & Smith 1992). We colonise the past and fashion our own fiefdoms. The past was (and still is) appropriated and contemporary 'others' become passive foreground to ancient landscapes, which are seen as open territory on which our logos becomes imposed. The generation of the 'truth' of the researcher with the 'scientific' approach or 'qualifications', and the power within the institutions of the dominant political, economic and academic systems becomes the dispatched truth. Worldviews that come to us through remaining oral traditions, material remains and/or representations of being/beingness that we term 'art' are thus reconstituted or annihilated in and through the language of the dominant culture of archaeology.

Meanwhile, concepts of social–cultural development can become linked with the discussion of humans as biological entities. Such 19th-century examples are seen in the work of Thomsen's tripartite categorisation of ancient tool technologies resulting in a relative chronology essentially based on the use of specific raw materials (Thomsen 1836). The Lower, Middle and Upper Paleolithic are another example of the evolution of tools tied to the evolution of biology and culture. Morgan's social evolution plan of human 'progress' moved from savagery to barbarianism (each divided into Lower, Middle, and Upper stages) to the pinnacle of civilisation (one stage only) with phonetic alphabet becoming a distinguishing factor for the civilised (Morgan 1877). Unfortunately, biological determinism was one of the results. Such ideas were current into the 20th century. Tylor, generally considered the founder of cultural anthropology, tells us to be wary of easily composed interpretations: '[with] writers on the Origin of Civilization being able to tell us all about it, with that beautiful ease and confidence which belong to the speculative philosopher, whose course is but little obstructed by facts' (1970, 137–138). Gould succinctly puts paid to the issue of representation from the perspective of the palaeontologist. His views easily translate to a critique of both archaeology and anthropology stating that 'Few scientists would view an image itself as intrinsically ideological in content ... [however, those] masquerading as neutral descriptions of nature ... are the most potent sources of conformity, since ideas passing as descriptions lead us to equate the tentative with the unambiguously factual' (1989, 28).

There has been a sea change, not quite a tsunami, in our understanding of our place in the grand scheme that Darwin proposed and the advancements made through the Neo-Darwinism debate. The most recent and, considerably, most cogent challenges to traditional and fairly recent stories of 'us' come from biological/psychological sciences – specifically DNA analysis and advances made in the study of cognition. And, of course, materiality plays here. Consider Lanza's First Principle: that which we perceive as a reality is 'a process that involves our consciousness' (Lanza 2009, 23). Here de Waal (2019) would challenge us to recognise that other biological beings are sensate as well and some leave evidence, however ephemeral, of their consciousness in material remains. Consider the use of sticks by chimpanzees for termite hunting and the selection of specific small dead branches on trees by common ravens (*Corvus corax*) when building a nest, a nest used by a multigenerational family over a number of years. But then, much of the conscious acts of all sensate beings are in the realm of the non-artefactual and non-symbolic deposition and additionally made from materials unable to survive the taphonomic interval. Surviving artefactualised evidence increased in number and complexity as we evolved to be the genus *Homo* and its subsequent and somewhat diverse species and sub-species. Actual hard ware remains! Therein is situated the interactional nature found in the whispers from the deep past that come to us essentially as shortcuts to our creating a unity with this past from what we can define as being alive to the ever-evolving present. Most evocative for us has been those things that seemed to address the complexity of mind and the qualities of feelings and not infrequently classified as objects of art. However, echoing down through time, material 'things' became our database to our seeking of self through time and space. But more: Simonetti (2017, 2) tells us there is a need to engage with

> real corporeal beings in their practical engagement with the world. ... [and their] relation to others ... fundamentally linked to an understanding of the body, perception and practice. Clearly, this required a fundamental shift in the view of the body as a mechanical entity, a perception that populates everyday discourse.

This supports any contemplation on the nature of our being sentient and agents of the configuration of the 'movable now' from and through the recovery of the compositions of near and far ancient pasts.

In recent years, however, we have had significant advances in the recovery of DNA from ancient times and this at a time where our understanding of the structure and function of the brain is enhanced by advances in DNA analysis, neurobiology and psychology. Beyond significant is the NOTCH family of genes, specifically the NOTCH 2 gene. It is a case in point. Approximately 8–14 million years ago there was a copying error that 'inserted' an extra non-functioning chunk of NOTCH 2 (consider *epigenetics*) (Henriques 2013; see also: https://www.psychologytoday.com/ca/basics/epigenetics). It is still found in gorillas and chimpanzees but not orangutans (they had parted ways some time pre-14 million years BP). About 3–4 million years ago, when we had parted from the apes, 'a second mutation activated the once non-functional code' and we have NOTCH 2NL (Tarlach 2019). It is present only in modern humans and, based on ancient DNA, our past near relatives, the Neanderthals and the Denisovans. What is important here to our understanding of the emergence of our cognitive abilities is that the NOTCH 2NL gene changed protein production resulting in neural stem cells becoming cortical neurons. This caused a significant increase in the number of neurons in the neocortex leading to larger and more powerful brains – brains with significant changes in cognitive functioning (Fiddes *et al.* 2018). Bigger brains come with a metabolic cost. González-Forero and Gardner developed a metabolic cost-benefit analysis considering three possible 'drivers' for brain evolution: social challenges (competitive/cooperative), cultural challenges (development of new skills and trans-generation learning) and/or ecological challenges (food acquisition and predator avoidance;

González-Forero & Gardner 2018). Their initial hypothesis was that social challenges would be the dominant driver for the development of brain/body ratio as found with the modern and near modern forms of the genus *Homo*. However, their results indicated that ecological challenges (60%), cultural challenges (30%) then social challenges (10%), in that order, created the best fit for the brain/body ratio that emerged for the modern and near modern forms of the genus *Homo*. They comment: 'our model indicates that brain expansion in Homo was driven by ecological rather than social challenges, and was perhaps *strongly promoted by culture* [emphasis added]'. For the anthropologist this is a gratifying result since all definitions of 'culture' state that it is learned behaviour, acquired as a member of a society and that it is trans-generational. And if we turn our eyes to what we term 'art' and our ears if we include the alliterative oratory of knowledge and the sign/signal/rhythm that could be generally defined as 'music',[3] by its nature it is a message, however transitory, for and to others that engages all in the *movable now* where the past becomes remembered and the future anticipated even as the voice of the elders becomes silent, lost in time.

This is not to say that we did not have a glimmering of advances that would allow the development of the ability to think of 'self' and to externalise our dreams. The increasing size of the brain became evident, and more so with geological time approaching us, with each palaeontological find of craniums of the genus *Homo*.

- *Australopithecus africanus*   461 cc
- *Homo habilis*   900 cc
- *Homo sapiens neanderthalensis*   1400 cc
- *Homo sapiens sapiens*   1300 cc

Still this is not the whole story of the brain. Size can be important, but only with the developing complexity of the structures of the brain itself. Consider Einstein with a cranial capacity below the statistical average for modern *Homo sapiens sapiens* and, pointedly, less volume than the average Neanderthal. However, there was significantly greater density of neurons than found in the average modern *Homo sapiens sapiens*. Organisation of the brain really matters! Casts illustrate size of the cranial vault and, to our great advantage in research, include the interior modelling of the skull created by the interaction between the maturation of both the bones of the skull and soft tissue of the brain as encased. It is interesting that each of the brain casts contributes to an understanding of cranial capacity and brain structure of a species but also show individual development. There are the discrete maps of a past individual. As such inter-species and intra-species comparisons can be generated. In the parlance of the archaeologist/palaeontologist both diachronic and synchronic variations can be charted and populations emerge. Nevertheless, the individual skull encased a brain that perhaps gave us some great work of the mind – such as the domestication of fire, the Levallois method of creating 'blanks' and increasing cutting edges. Merlin Donald, as we were on the dawn of the detailed DNA work of today, did give us an ideational map of '… three major cognitive transformations by which the modern human mind emerged over several million years, starting with a complex of skills presumably resembling those of the chimpanzee...' (Donald 1991, 1) that changed in content and complexity. Donald proposed a three-stage development of human symbolic capacity through culture (Donald 1997; **emphasis added in bold**):

- Mimetic culture: The watershed adaptation allowing humans to function as symbolic and cultural beings was a revolutionary **improvement in motor control**, the '*mimetic* skill' required to **rehearse** and refine the body's movements in a voluntary and systematic way, to **remember** those rehearsals, and to **reproduce** them on command. Following this development, *Homo erectus* **assimilated** and **reconceptualised events** to **create** various **pre-linguistic symbolic traditions** such as rituals, dance, and craft.
- Mythic cultures arose as a result of the acquisition of speech and the invention of symbols. Mimetic representation serves as a pre-adaptation to this development.
- Technology-supported culture: Finally, the cognitive ecology dominated by ephemeral face-to-face communication has changed for most of us as a result of the **external memory-store** that reading and writing permit. Computer technology intensifies these changes by offering even more extensive capacities for external storage and retrieval of information.

A brain increase from 300 to 1300 cc has not necessarily been a journey in a linear fashion (Gould 1997) and the upshot is that we are not yet done. Indeed, we are a work in progress. Many agree with Donald that externalised memory-storage has costs. We are only starting to study, in depth, the consequences of memory by proxy. Is it possible that memory is found at the somatic level and is best understood through phenomenology – the study of consciousness as it emerges from the interactional states we experience? Our understanding of this is important 'because the human cognitive system, *down to the level of its internal modular organization*, is affected not only by its genetic inheritance, but also by its own peculiar cultural history' (Donald 1997, 362). Most of us have had an awakening – a moment when things seem to *pop* to mind, into conscious view creating a new understanding. Awakening consciousness becomes a done deal that has its own 'now' that there is no way back from – once realised so be it! This, step by step, brings us to oral traditions with a move from signal to symbol more or less translated to language and artefactual representations (art, tool, etc), then, for some, to a literate form substantiated first through representative symbols (writing) eventually intermediated by binary digital coding. The good old archaeological term for the order of getting things done, *chaîne opératoire*, describes our stages of developing consciousness – it had to sequence. With *art*efact as 'text' we may find indicators of the various cognitive thresholds reach by our genus. Our speculations need to focus on the potential triggers for cognitive shifts and the possible serendipitous flashes that can create such 'a-ha' moments. We can seek this in our own reflexivity and the more so if we accept that reflexivity, in the past, was an initiating point for analysis of being even in deep time.

For so long we have been of two minds on our ancestry. And, of course, recent DNA studies have shown that we really were and still are of two minds – one that pre-dated the advent of the genus *Homo* (*erectus* 'varieties' here included) and one that could well be considered the actual marker of the 'modernity' of our genus. Now we alone carry forward today the construct of a modernity once shared with at least two[4] other *Homo* forms in the relatively not so very distant past – Neanderthals and Denisovans. Why is this important? Here we could give a belated recognition of Jaynes and his concept of the bicameral mind (Jaynes 2000) that currently is, in certain aspects of analysis, supported by the work of McGilchrist (2009) with his emphasis on hemispheric brain functioning from ancient times to the post-modern era of our 'now'. His work, and that of his contemporary colleagues, has shown the increasing dominance of the left brain (methodical, analytical, and linear). McGilchrist (http://frontierpsychiatrist. co.uk/interview-with-iain-mcgilchrist/) continues by placing emphasis on

> Some very subtle research by David McNeill, amongst others, [that] confirms that thought originates in the right hemisphere, is processed for expression in speech by the left hemisphere, and the meaning integrated again by the right (which alone understands the overall meaning of a complex utterance, taking everything into account)

The right brain – the creative hemisphere – is the symbol maker capable of metaphoric thought. It is this aptitude that moves us to what some cultures call 'art' while others do not work within such terminology rather seeing it as the capacity to move into the consideration of the transcendental, the presentation of alternative forms of reality embracing life's essential ambiguity, or the depiction of the life lived in dynamic, sometimes contradictory ways. This takes back to the beginning the model/ metaphor found in the ***qbit***. Reflect on the fact that pre-*Homo* and very early *Homo* forms left material culture items – specifically stone tools, considered by some as things of beauty and indicators of survival and change, however slow in our specific evolutionary journey. The bind here is that we have little indications of the ephemeral mediums for the generation of what could be defined as representations of the meditative mind. But in a reflective moment when something has come from their hand to yours, from the dusts of time (perhaps the dust of their physical being), and you have seen/touched perhaps a bulb of percussion, the scar of a carefully removed flake, held their tool in your hand ... is there not the sense of a contemplative mind at work? Recent ethnographic examples can display this contemplative, and with reflexivity, shared space – this intersubjective moment. Specifically, here, Navaho Sand Paintings[5] are acts of, what we could call for want of a more culturally precise term, 'worship' that incorporate observations of the natural world, mythic symbols, and representation of relationships with powerful transcendental beings. Many have directional orientation (north, south, east and west) with associated imagery including both humans and animals and symbols of such things as rainbows. The use of specific colours in the Navaho Sand Paintings are also widely seen in the art of other cultures and so too in their earliest forms – white, black and red (sometimes brownish in display) use is very ancient and then there was the addition of blue and yellow in some cultural representations.

The emergence of the metaphorical 'self' – and the poetic imagination – is evidenced in what will eventually be called those 'kinds of beliefs and rituals we label "religious" [that] are so tightly interwoven into … everyday thought and action that [there is] … no word for them' (Wright 2009, 20). And our sense of awe in beingness shines forth. I do not make an equivalency between the spiritually based paintings of the Navaho and what could have been scribed in the sands of the very, very ancient and even pre-genus *Homo* past. No such statement is here inferred as is so frequently seen in the era of the discussion of 'primitive' art. The point here is the fragility of such creations. For art such as sand paintings there is, in essence, no possibility of survival of such acts since a hand brushed through, a rainstorm, a wind making dust and all is gone. It is so exceedingly rare to find the accidental survival of being in the ancient 'now'. The most famous examples are the Laetoli pre-*Homo* footprints, showing early bipedalism frozen in time by accident of a soft rain, found by Mary Leakey in 1976. Sadly, we do not know if the glimmerings of the mind of symbolic representations were active in the very early stages of our evolution since we do not have the transient, fleeting data of the metaphorical as composed of materials susceptible physical taphonomy. We have more from the lithics for us to learn from in the sensing of the mind but perhaps a mind based in the pragmatics of the day. Written in stone, taphonomy has left such artefacts untouched until our hand reclaims them.

So, what is there, for example, with early stone tools? They give us the fracture mechanics of the day – the physics of their modern times, the recognition and acquisition of appropriate materials – perhaps through exchange with others, the 'school-room' training at a debitage location, and we can go onward from here to use, refitting and discard … etc (Dods 2017). Somewhere some senescent being looked and recognised an advantage from a specific, perhaps pragmatic, act and this act became generalised to a wider population. Of course, we do have similar examples of this with other primates. The 'Snow Monkeys' of Japan (*Macaca fuscata*) use the hot springs to weather the storms and temperature of winter. This was their 'invention' and it became a learned activity through their families and troops. Japanese macaques engaged in what was termed 'pre-cultural' activities in the washing of food in the sea (Visalberghi & Fragaszy 1990). However, the most persuasive articles on primates and culture come from the works of de Waal (1999), Whiten (Whiten *et al.* 1999) and, earlier, Van Lawick-Goodall (1973). Subsequently, de Waal gives us an understanding of the significance of *mirror neurons* in that they 'don't distinguish between our own behaviour and that of someone else, so they allow one individual to get under another's skin … [this emphasizes] the profound implications for imitation and other forms for bodily fusion'. Important as this is to understanding our acquisition of culture, de Waal continues by emphasising that this discovery came to us through the study of the brains of macaques and 'monkey-see, monkey-do' behaviour. It is not an accidental observation or use of a *turn of phrase* since 'Primates are natural conformists.

Not only do they imitate, they also like to be imitated ... [and of course] we regard imitation as a compliment' (de Waal 2019, 94–95). To imitate, in certain situations, has been equated with obsequiousness, flattery, fawning – all in the service of being in the right group – troop and understanding the right metaphor, or trope.

The consideration of becoming engaged in the intersubjective

> ... if you wish to ask the question of the ages-why do humans exist ... [it is] because *Pikaia* survived the Burgess decimation ... We are the offspring of history, and must establish our own paths in this most diverse and interesting of conceivable universes – one indifferent to our suffering, and therefore offering us maximal freedom to thrive, or to fail, in our own chosen way (Gould 1989, 323).

To reiterate, we live in an ever moving 'now', a 'here' that is constantly changing and changing us, even when we experience the sensation that time stands still. The understanding of time-consciousness is core to any investigation of the 'meaning' of our emerging awareness and the artefactual remains of our becoming aware. In the accident of being and the understanding self-ness we become situated in activity and capacity and a sense of potentiality (Kimhi 2018, 69–70).

Gallagher comments: 'perceiving succession and change would be impossible if consciousness gave us merely a pure momentary slice or if the stream of consciousness were a series of unconnected experiential points ... [merely a] succession of isolated, punctual, conscious states does not add up to a consciousness of succession and duration' (Gallagher 2019; see also Open Peer Commentaries and author,[6] 98–116). Consequently, we need to be careful not to focus on the 'object' without considering its place in an enactive integration. The object is the deep complex representation of a *qbit* contemplation. Succession/duration are core to working through the layers so we can speak of the diachronic and the synchronic thereby making time/space linkages and meanings. These linkages awaken us to the fact that the past is not static but representing/presenting to us a frequently unsuspected dynamic (Dods 2015). Here is an example through consideration of cave painting such as those of Lascaux. Season after season the animals in all their abundance and lively representation were painted. New images merely reinforced the representation of their abundance. Accompanying such images with traditional oratorial presentations of the lives lived (human and other than human) creates *storia*, each composed of multiple layers of embedded meaning arising from the oral in sympathy with that 'written' appearing as pictures. Meditative reflection can be an act of consecration, commitment, one that could be carried in the mind through remembering the imagery long after the spoken accompaniment fades from deep, specific, remembrance. Christopher de Hamel speaks of this when he considers ancient manuscripts. Here multiple levels of meaning, fundamentally *lectio divia*, progress to sincere, perhaps devout, reflections based on the mind's eye remembering then reflecting or dreaming on the dynamics of the art on the created decorated surface (de Hamel 2017, 224). Do not let the wall surface faze you – it has been transformed and is metaphorically layered thereby most certainly multi-dimensional. We can be in sympathy with such a response as they have shown us. This challenges us to seek their story, see their lives. Now think of walking through such Upper Palaeolithic halls by torchlight where from the darkness emerges liveliness. The flicker of the light gives the images a seemed motion and a revitalization of the lives we can yet see from that distant past. They are, in essence, motion pictures – not of the pixel world we live in but a world of magic and mystery and wonder and the awe of becoming – all notions of an animated, spirited world. Our core question arises from our wonderment in considering when in our biological and socio-cultural development did the genus *Homo* make such a remarkable transition?

It is from the dynamic of time and our composed integration of what we consider a primary past that we compose our hoped-for futures. The Who, What, Where and When augmented by the How causes us to question the coming together of the bits and pieces of the past that allow us to compose an explanation for our transition into conscious interaction with our natural selves and natural environments. Essentially, we seek to understand our foundaetional past – a past where we became gradually aware of our positionality in time and place, a past that gives us pace in our family of time. Some of this comes to us in evidence of our defining of what we now call cultural landscapes and through this we start to see the defining of ourselves – the naming and nurturing of place. In a sense we are seeking a historical perspective on the development of the consideration of Being. In the examination of the past, the 'there'/'then', objects we call artefacts and sometimes attribute to an artistic sensibility, are the illustration of thought and thus they provide a framework of the context and structure of what anthropologists call a *Worldview*. However fundamental it may seem to us in its initial composition it is still a picture of a knowing of a place in the cosmos.

Nevertheless, for deep time cultural analysis there exists an imbalance in the emic/etic relationship because frequently only the artefact stands for the informing of and thus the development of an intersubjective space. The researcher, situated in her/his own cultural reality, gives meaning to the 'now' of the 'then' without the development of a negotiated meaning. In same-time, face-to-face negotiations of meaning, intersubjective space is malleable, consultative – somewhat, more-or-less, sometimes. This is because it shares intrinsic qualities of the movable 'now'. Thus, there can emerge oxymora to qualify paradoxes and illustrate contradictions. Time exposes alternative explanations of the meaning of the same items, although structured initially via the medium of science, since the interpretation comes through the cognitive ever-changing world of the researcher. We have physical objects – artefacts that are products of the human mind and the subjective experience of that 'human' mind of the past from which the composition of an object was imagined and then materially realised. And then there is the subjective experience of the researcher defining an '*arte*fact' (Popper 1978) – essentially engaging in the process of reverse engineering to reach an understanding of the mind of the maker. This is an attempt to compose a meeting of minds – the intersubjective space that the anthropologist enjoys, sometimes, if they understand what they are attempting to accomplish in a context devoid of a narcissistic smugness of intellectual elitism. Intersubjective space is created by understanding the world of *savoir* as defined by Lyotard, where knowledge is not located in a set of denotive statements (the world of *connaissance*) but in a knowledge that holds the 'how to' enacted in the competence of life (Lyotard 1987, 78–79). Here is the 生き甲斐 (*ikigai*) introduced earlier. The

> application of criteria of efficiency (technological qualification), of justice and/or happiness (ethical wisdom), of the beauty of a sound or colour (auditory and visual sensibility) ... The consensus that permits such knowledge to be circumscribed and makes it possible to distinguish one who knows from one who doesn't (the foreigner, the child) is what constitutes the culture of a people (Lyotard 1987, 78–79).

Artefacts and what we call 'art' charm but context informs interpretation. Landscapes, natural choices and cultural constructs, speak to our emerging humanity and as Lessard tells us they are a 'mirror of ourselves, then and now' (Lessard 2019). Landscapes entwined with the evolution of our biological selves have brought us to the 'now' that gives us cognitive skills to engage in wondering about and being enchanted by the past while contemplating the future. Now, to a look in the mirror to see ourselves and the wonders of what we have wrought – hopefully with humility.

Archaeologists' work is chancy – right place, right time, right funding, right ..., etc. It is the chance of it all that generates the *frisson* – essentially a gambler's world of random positive/negative reinforcement much like an ancient hunt for a mammoth, except that was life or death. Some of the reinforcement comes from early material-based memories through the lens of the culturally constructed and technologically expanded categories that we consider important to the framing of our understanding. Theory and methodology range from research focused on our biological evolution to material analysis progressing to the ascetic musings of *art* as we understand it. Relatively recent advances in analysis in many disciplines inform us to challenge our current and sometimes long-held assumptions. However, new information from the realms of science (DNA analysis, changing concepts of time and space, emerging understanding of the evolution of consciousness as it operated/operates in the human species, work on cognition, etc) move us to consider and attempt to incorporate newer and deeper considerations of the meanings we have traditionally attributed to the products of archaeology.

There has been hubris in our claiming of the civilised slot. Not infrequently, we have situated ourselves in the position of power and the ownership of the interpretation of 'others', ancient and recent. We, for the most part, have lived in a world, or class, of specific advantage that allowed us to claim the prerogative to define the *here and now* and the *there and then*. To be reasonable, there is much work that is value free and offers us observations that are generally considered to be authentic to the data. For researchers who have superseded their cultural biases kudos should become mandatory. Then there is the work that results in inappropriate make-believe that damages our ability to engage in the power of the imagination and respect the magic of reflexivity. The problem with the omnipresent media dissemination of 'alternative facts' (here come the Mayan, Egyptian, Incan, *etc.*, aliens, again) is that alternative facts become accepted by the gullible as truths. How do we move beyond self-defined authority to decolonise archaeology while at the same time embracing and supporting that which science (magic in the past) can illuminate? Make-believe may be 'fun' but it does not give us access to the grandeur of our journey of becoming. It is difficult, but not impossible, to aim for the development of an inclusive, intersubjective space mediated by challenging our own worldviews. This, effectively, is what Ingold terms the 'act of remembrance' and the course 'of engaging perceptually' with a world 'pregnant with the past' (Ingold 1993). This means, distinct from quantification, *entering* the analysis and engaging in ways that may seem unscientific, as this approach is not of the world of *connaissance* but rather of the world of *savoir* (Lyotard 1987). In challenging the researcher to become embedded in the world of the past there is the potential to create an alternative response to the artefact, and too what we call *art*. This is a response removed from current fallback positions of 'it must be ritual' although ritual must not be excluded as a possible answer, just not the only answer, or the simple answer. 'The caveat here is that we must recognize that so many cultures regard all that they do as part of the sacred there being no separation of the sacred from the profane, no oppositional statements about "being"' (Dods 2007). Here is the both/and, again. Indeed, the analysis generated could be considered to be in the realm of *quietism*, an acceptance of how things are and the personal decision not to change or alter their beingness (Kimhi 2018, 159–161). Some of the insights we come to may well be syncategorematic in form – they only come to a meaning if and when we can situate that insight into a wider realm of knowledge as constructed in a specific society or culture. Because of the vagaries of taphonomy (actual and metaphorical) knowledge is 'partitioned' and some of these partitioned off spaces operate much like the information encased in the secret societies found in many cultures of today. We are 'other' to their being and becoming and thus not given *entrée*.

Provocative in the context of 'otherness' reiterated is a comment from Caleb Scharf (2017) in his discussion of the observable universe. He makes a distinction between atoms that have cycled through sensate beings and those that have not. He does not make the distinction that the cycling through *Homo sapiens sapiens* specifically supersedes that through all the genus *Homo* or any other form of *life*. However, in accepting that the state of being sensate affords us the ability to 'experience that slippery property known as consciousness' (Scharf 2017, 74) we can extrapolate the meaning of being as seen in the gifts from the past that we term archaeological *data*, all of it including pre-literate society, materially based surviving the physical and metaphorical taphonomy mentioned above (Dods 2017). Indeed, Lanza extended this approach even further, noting:

> Our science to date has failed to recognize those special properties of life that make it fundamental to material reality. This view of the world in which life and consciousness are the bottom line in understanding the larger universe – biocentrism – revolves around the way a subjective experience, which we call consciousness, relates to a physical process (Lanza 2009, 13).

When did we become conscious of being in a sentient state? When did we come to the capacity to dream of a metaphoric external world? When did our dreams become externalised common, shared themes? And what do the manifestation of these themes tell us about our understanding of beingness shared with others, our emerging shared worldviews? We ask: tell us your story and bring us to an intersubjective space that comes back to us as your dreams now manifest in our understanding. The gift we received with our evolution is reflexivity. Never mind fun with Schrödinger's cat. Did it have a hat? Rather, think on the essential meaning and the expansive questions from Dirac's equation $(i\,ɤ\cdot\partial\psi = m\psi\,ɤ)$[7], again the metaphoric **qbit** but also the opening to so many questions/answers on being and the cosmos. When unpacked (deconstructed?) it opens layers on layers of meaning between nothingness and the infinite. There are meanings to being sentient and the levels of our emerging humanness. Some of this can be puzzled through our representations of our place in the world, indeed in the universe. Much of this is beyond metaphor. It is a challenge to us, a challenge to seek depth, to see meaning and to contemplate our emergence of understanding. *Cherchez.*

Notes

1. http://maryarrchie.com/2019/01/21/vedbaek-burial-a-baby-buried-upon-swans-wing/
2. A musical surprise: https://www.youtube.com/watch?v=K5626WzsfMw
3. Egyptian drumming https://www.youtube.com/watch?v=0rvPpKW9Zz8 and Whistling/sounds https://www.youtube.com/watch?v=oVinnWp1Pck
4. The recent find in the Philippines of *Homo luzonensis*, dating to 50,000–67,000 years ago, suggests a third pre-*H. sapiens sapiens* species. (*Nature,* April 2019) https://www.nationalgeographic.com/science/2019/04/new-species-ancient-human-discovered-luzon-philippines-homo-luzonensis/
5. Navaho Sand Paintings https://www.crystalinks.com/navajo.html
6. https://constructivist.info/13/1/091.gallagher.pdf
7. https://howiefirth.files.wordpress.com/2012/07/cerndirac3_9-02.jpg; https://www.youtube.com/watch?v=tCIHcX5-mq0

References

Ball, M.S. & Smith, G.W.H. 1992. Analyzing visual data. *Qualitative Research Methods* 24. Newbury Park CA: Sage.

Bello, S.M., Delbarre, G., Parfitt, S.A., Currant, A.P., Kruszynski, R. & Stringer, C.B. 2013. Lost and found: the remarkable curatorial history of one of the earliest discoveries of Palaeolithic portable art. *Antiquity* 87(335), 237–244.

Campbell, J., Moyers, B. & Flowers, B.S. (eds) 1988. *The Power of Myth.* New York: Doubleday.

de Hamel, C. 2017. *Meetings with Remarkable Manuscripts.* New York: Penguin Press.

de Waal, F.B.M. 1999. Cultural primatology comes of age. *Nature* 399(6737), 635–636.

de Waal, F. 2019. *Mama's Last Hug.* New York: Norton.

Dods, R.R. 2004. Knowing ways/ways of knowing: Reconciling science and tradition. *World Archaeology* 36(4) Debates in World Archaeology, 547–557.

Dods, R.R. 2007. Intersubjectivity and the meaning of things. *International Journal of the Humanities.* Annual Review 4(9), 99–106. doi: 10.18848/1447-9508/CGP/v4i09/58244.

Dods, R.R. 2015. Seeking the mind of the maker. In D. Gheorghiu & P. Bouissac (eds) *How Do We Imagine the Past? On Metaphorical Thought, Experientiality and Imagination in Archaeology.* Newcastle upon Tyne: Cambridge Scholars Publishing, 9–26.

Dods, R.R. 2017. The never ending journey: Cycling and recycling seen through a critical assessment of the taphonomic process. In D. Gheorghiu & P. Mason (eds) *Working with the Past: Towards an Archaeology of Recycling.* Oxford: Archaeopress, 1–19.

Donald, M. 1997. The mind considered from a historical perspective: human cognitive phylogenesis and the possibility of continuing cognitive evolution. In D. Johnson & C. Ermeling (eds) *The Future of the Cognitive Revolution.* New York: Oxford University Press, 478–492.

Donald, M. 1991. *Precis of Origins of the Modern Mind: Three stages in the evolution of culture and cognition.* Cambridge MA and London: Harvard University Press.

Errington, S. 1998. *The Death of Authentic Primitive Art and Other Tales of Progress.* Berkeley: University of California Press.

Fiddes, I.T., Lodewijk, G.A., Salama, S.R., Jacobs, F.M.J. & Haussler, D. 2018. Human-specific *NOTCH2NL* genes affect notch signalling and cortical neurogenesis. *Cell* 173(6), 1356–1369.

Gallagher, S. 2019. The past, present and future of time-consciousness: From Husserl to Varela and beyond. *Constructivist Foundations* 13(1), 91–97.

González-Forero, M. & Gardner, A. 2018. Inference of ecological and social drivers of human brain-size evolution. *Nature* 557(7706), 554–557. doi: 10.1038/s41586-018-0127-x. Epub 2018 May 23.

Gould, S.J. 1989. *Wonderful Life: The Burgess Shale and the nature of history.* New York: Norton.

Gould, S.J. 1997. Unusual unity. *Nature* 106(3), 20–23, 69–71.

Henriques, G. 2013. Theory of knowledge. A revolution in evolution: A return to Lamarck? *Psychology Today.* https://www.psychologytoday.com/ca/blog/theory-knowledge/201312/revolution-in-evolution-return-lamarck

Hollander, L.M. (ed.) 1962. *The Poetic Edda: Translated with an introduction and explanatory notes* (2nd ed., rev. ed.). Austin TX: University of Texas Press.

Ingold, T. 1993. The temporality of landscapes. *World Archaeology* 25, 152–174.

Jaynes, J. 2000. *The Origin of Consciousness in the Breakdown of the Bicameral Mind.* Boston MA: Houghton.

Jung, C. 1933. *Modern Man in Search of a Soul.* London: Kegan Paul, Trench, Trubner and Co.

Kimhi, I. 2018. *Thinking and Being.* Cambridge MA: Harvard University Press.

Lanza, R. 2009. *Biocentrism. How Life and Consciousness are the Keys to Understanding the True Nature of the Universe.* Dallas TX: Benbella Books.

Lessard, S. 2019. *The Absent Hand: Reimagining our American landscape.* Berkeley CA: Counterpoint Press.

Lyotard, J.-F. 1987. The postmodern condition. In K. Baynes, J. Bohman & T. McCarthy (eds) *After Philosophy: End or Transformation?* Cambridge MA: MIT Press, 73–94.

McGilchrist, I. 2009. *The Master and His Emissary: The divided brain and the making of the western world.* New Haven CO: Yale University Press.

Miller, A.L. 1987. The Swan-Maiden revisited: Religious significance of "divine-wife" folktales with special reference to Japan. *Asian Folklore Studies* 46(1), 55–86.

Morgan, L.H. 1877. *Ancient Society: Researches in the lines of human progress from savagery through barbarism to civilization.* New York: Holt.

Popper, K. 1978. *Three Worlds.* The Tanner Lecture on Human Values. University of Michigan, April 7, 1978. https://tannerlectures.utah.edu/_documents/a-to-z/p/popper80.pdf

Renfrew, C. 2004. Towards a theory of material engagement. In I.N.E. DeMarrais, C. Gosden, & C. Renfrew (eds) *Rethinking Materiality: The engagement of mind with the material world*. Cambridge: McDonald Institute for Archaeological Research, 23–32.

Robertson, R. 2011. *Grave Goods. The Wrecking Light: Poems*. New York: Mariner Books, Houghton Mifflin Harcourt.

Rosenberg, D. 2007–2008. *Marking Time*. http://cabinetmagazine.org/issues/28/rosenberg.php

Scharf, C.A. (illustrations by Ron Miller) 2017. *The Zoomable Universe: An epic tour through cosmic scale, from almost everything to nearly nothing*. New York: Scientific American/Farrar, Straus & Giroux.

Schumacher, B. 1995. Quantum coding. *Physical Review* A51, 2747.

Simonetti, C. 2017. *Sentient Conceptualisations: Feeling and thinking in the scientific understanding of time*. London: Routledge.

Tarlach, G. 2019. Human origins: How to build a brain. *Discovery* 40(1), 40–41.

Thomsen, C.J. 1836. *Ledetraad til Nordisk Oldkundskab* (Guide to Northern Antiquity), published in English in 1848.

Tylor, E.B. 1970. *The Origins of Culture* (originally published as Chapters I–X of *Primitive Culture* by J. Murray, London). Gloucester MA: Peter Smith.

Van Lawick-Goodall, J. 1973. Cultural elements in a chimpanzee community. In E.W. Menzel (ed.) *Precultural Primate Behavior*. New York: Basel, 144–184.

Van Reybrouck, D. 1998. Imaging and imagining the Neanderthal: The role of technical drawings in archaeology. *Antiquity* 72(275), 56–64.

Varela, F.J., Thompson, E. & Rosch, E. 1992. *The Embodied Mind: Cognitive science and human experience*. Cambridge MA: MIT Press.

Visalberghi, E. & Fragaszy, D.M. 1990. Food-washing behaviour in tufted capuchin monkeys, *Cebus apella*, and crabeating macaques, *Macaca fascicularis*. *Animal Behaviour* 40(5), 829–836.

Whiten, A., Goodall, J., McGrew, W.C., Nishida, T., Reynolds, V., Sugiyama, Y., Tutin, C.E.G., Wrangham, R.W., & Boesch, C. 1999. Cultures in chimpanzees. *Nature* 399(6737), 682–685.

Wright, R. 2009. *The Evolution of God*. New York: Little Brown and Company.

Chapter 2

Catching the ephemeral – aesthetics of artful artefacts

A Middle Stone Age Still Bay bifacial pointed stone tool from Blombos Cave, South Africa and a Migration Period brooch from Kvåle in Sogn, Norway

Torill Christine Lindstrøm

'Aren't they beautiful?' – 'Are they works of art?' – These questions strike us as we find, analyse, observe, and experience archaeological objects – 'things', 'artefacts' from ancient times. Because, yes, while observing and analysing, we 'experience' also. With our minds and emotions. And our *personal* experiences influence our *scientific* observations and analyses in subtle ways, consciously, sub-consciously and un-consciously. Part of the 'experience' is what we estimate as 'art' and evaluate as 'beautiful' and 'aesthetic'. In this chapter, I endeavour to use a method for evaluation of aesthetic characteristics of objects; and I will try to find out whether archaeological artefacts, at all, can be tested regarding aesthetic properties.

First, I will present some perspectives on aesthetics. Next, I will present research regarding establishing objective criteria for subjective evaluations of aesthetics. Then, I will take this set of criteria regarding aesthetics and use it as a filter on my experience and analysis of two selected archaeological artefacts from completely different times, cultures and contexts. Finally, I will discuss whether it is probable that they were made with the intention of creating them as aesthetic, artful artefacts. (Note: I will tend to use the words 'aesthetics' and 'beauty', and the words 'artefact' and 'object' interchangeably).

De gustibus non est disputandum

… said the ancient Romans. It means: 'Taste can not be discussed'. In this proverb they probably included all kinds of 'taste' and preferences, but also included aesthetics in general: that which is beautiful and pleasant to perceive. This proverb expresses a *subjectivist position* to aesthetics; each subject/person has his or her own view. Also other cultures have similar proverbs, like the Norwegian: *Smak og behag kan ikke diskuteres* ('Taste and pleasure can not be discussed'), and the rougher one: *Smaken er som baken, delt i to*, ('Taste is like the bottom, divided in two'). And then there is the English: 'Beauty is in the eye of the beholder'. In short, what all these proverbs express is the notion that opinions differ regarding aesthetics, and that's it.

– Is it?

In contrast, from the times of classical Greece, theories expressing an *objectivist position* about aesthetics have been proposed. The Greeks claimed that aesthetics depended on κοσμετικοσ (*kosmetikos*), meaning balance and proportions. For one thing, it meant that there should be no exaggerations. The modern world's sex-bombs would probably not be considered beautiful by the ancient Greeks. I think they would have said: 'The breasts, bottoms and hips are too big and the waists are too narrow!' Their love goddess, Aphrodite, they sculpted with far more moderate forms. In the Renaissance, Leonardo da Vinci continued the Greek idea of proportions into his meticulous analysis of the proportions of the human body, and used it in his art. And, needless to say, his art was, and still is, a central force in European art and in European ideas of aesthetics and beauty.

What characterises these ideas of 'balance' and 'proportions' is that both those concepts pertain to characteristics of the *object*. The object: the human body, the human face, the body of an animal, the landscape, the sculpture, the vase, the jewellery, the painting, etc, etc, *has, or has not*, balance and proportions. These characteristics of beauty and aesthetics are defined as objective facts connected to the object. It is present *there*, in, on, and within the *object*, irrespective of the viewer and the viewer's opinion and taste. Regarding theory of science, one could say that this idea is an example of an objectivist position. Such a reliance on objective, observable facts is, within theory of science, called Structuralism, and within archaeology: Processualism.

Within this perspective, even very old psychological experimental research has contributed to the identifications of critical characteristics of aesthetics in objects. Not unexpectedly, *balance and proportion* came up as essential features (Fechner 1876; Birkhoff 1933; Arnheim 1974; Gombrich 2014), along with *symmetry* (Birkhoff 1933; Arnheim 1974; Gombrich 1984; Humphrey 1997). Symmetry is actually often an integral aspect of balance and proportion, so that these three factors are compatible, and relatively easy to identify. Additional factors are *informational content and complexity* (Eysenck 1941; Berlyne 1971; 1974; Garner 1974); and *contrast and clarity* (Gombrich 1984; Solso 1997). Informational content and complexity are related. The more the stimulus complexity increases, the more the amount of content and information in each object increases.

This objectivist position has a counterpart, an opposite position; the subjectivist position, in theory of science often termed Social Constructivism, and Post-Processualism within archaeology. From this perspective, the experience of aesthetics is a highly subjective event, connected to the history and preferences of each individual perceiving subject, but also to the individuality of each culture: different cultures (and sub-cultures) have their own idiosyncratic and subjective views about aesthetics. The subjectivist position therefore argues that any discussion about trying to identify common characteristics and criteria connected to aesthetics and 'beauty' is in vain. Simply: *De gustibus non est disputandum.*

And indeed, this position has a point. For instance, when looking at objects from various regions of the word, in particular everything *traditional* – costumes, art, architecture, and gardens – one must admit that they are different. Very different. Yet, somehow, persons from one region of the world are perfectly able to appreciate, applaud and enjoy such objects (costumes, art, architecture, gardens) from completely different regions and cultures than their own. This observation is striking. Which position is right then? Are there objective criteria for aesthetics or are aesthetic experiences subjective and idiosyncratic? Or could there be a third position?

Plato did it again!

Plato (and later theorists following him) expressed this idea: beauty is a property of an object that produces a pleasurable experience in any suitable perceiver (Tatarkiewicz 1970). Here Plato goes beyond identifying the *objective* features of an aesthetic object and demands the incorporation of *both the object and the subject* perceiving it. Thus Plato 'did it again'; he anticipated the third position within theory of science in psychology: *interactionism* (Ekehammar 1974). Whereas *personologism* explains persons and their behaviour primarily from factors *within* the persons and their personal life-history, and *situationism* explains persons primarily from factors in the *context* around the persons, contrastingly, *interactionism* claims that both factors within persons and factors surrounding a person, contribute, in interactive ways, to the characteristics and behaviours of a person. This pertains also to each person's ways of perceiving, experiencing and imagining. A parallel to this interactionism in psychology is epigenetics within biology and medicine; hereditary given genes (within the person) are 'turned on' and 'turned off' depending on contextual factors (surrounding the person). In a similar way, regarding the evaluation of beauty and aesthetics, from an *interactionist* position, one would claim that both the *objective* features of objects, *and* the *subjective* features of the perceiving subjects (persons), would interact during an evaluation of an object. Plato's idea contains exactly these two interactive elements: the objective and the subjective. But is Plato's interactionist idea applicable for aesthetic evaluations?

Regarding properties of the *object*, he is likely to have thought of the Greek criterion of *kosmetikos*: balance and proportions. Regarding the perceiving *subject*, he is rather specific regarding the qualifications: the perceiver (the subject/the person) should experience *pleasure*, and the perceiver should be *suitable*. What may he have meant by this? Regarding 'experiencing pleasure', I assume he would have considered any emotion of a hedonic tone. It is harder to fathom what he meant by '*suitable perceiver*', but a reasonable guess is: anybody who is able to perceive in a basic way (see, hear, feel, experience) the object in question (painting, sculpture, music, etc). But he might also have meant those who are sufficiently informed and educated to be appreciative of the aesthetics of objects, and had experience with them. Nevertheless, it is interesting that Plato includes qualities of, and within, the perceiving subject into his definition of beauty and aesthetics. This is indeed a very modern interactionist position; that it is the interplay between factors in both object and subject that creates the aesthetic experience.

Plato's interactionism meets the interactionism of today

Rolf Reber and collaborators conducted a series of experiments regarding aesthetics and art appreciation (Reber *et al.* 1998; 2004; Reber & Schwarz 1999; 2001; Reber 2002), from the perspective of cognitive psychology, summed up their findings and related them to other works and perspectives (Reber *et al.* 2004). They concluded that, regarding *objects*, the aesthetics of an object is dependent on *figural goodness (gestalt, from German:* Gestalt*), figure-ground contrast, stimulus repetition, symmetry, and clarity*. Gestalt is probably the most important feature of an object. Gestalt literally means 'form', but is as a cognitive concept often used to mean 'goodness of form'. It is also used to define an object *as* an object, and as a perceivable object. And, when the gestalt of an object is good, the object, 'objectively', has aesthetic properties. *Contrast* means that an object stands out as 'a figure', a separate 'thing' against 'the background' of the surrounding stimulus situation. *Stimulus repetition* means that a repeated stimulus is regarded as more aesthetic than if it stands alone, or is presented only once. This means that a compilation of several similar, or same, objects will be regarded as more aesthetic than one of these objects alone. *Symmetry* simply means that symmetrical objects are regarded as more beautiful and aesthetic than non-symmetrical ones. *Clarity* is a more complex concept, but means that if an object has a clearly defined structure that makes it easy to perceive, it is likely to be regarded as aesthetic.

However, Reber, Schwarz & Winkielman (2004) also discovered, or one might say, disclosed, important properties regarding the perceiving subject during aesthetic evaluations. They identified certain characteristics of the subject that pertained to the ability to perceive an object as aesthetic. One might say that these are properties of the *suitable perceiver* as Plato phrased it. Reber, Schwarz & Winkielman (2004) listed these characteristics as being: *experience with stimuli, implicit learning, prototypicality of stimuli, simplicity/complexity*.

This means that in order to be able to experience the aesthetic properties of an object, the perceiving subject should have had *experience* with the object and its stimuli. This means that the subject has had repeated exposures with it. Consequently, the subject would be able to recognise it among other objects. This experience would thus also give *implicit learning*, which means that the rules pertaining to the object would be unconsciously grasped and understood. The *prototypicality* of the object is another important factor. It means that the subject regards the object as a typical example of its class. This emerges from the interface between the subject, the object and similar objects, but is defined and evaluated by the perceiving subject, and therefore is more connected to the subject than to the perceived object. Similarly, whether an object is regarded as *simple* or *complex*, is dependent on the subject's evaluation, and of course, related to the subject's experience with the object, and with the subject's cognitive abilities. A simple to medium level of complexity is usually the preferred degree, whereas very simple and very complex objects (or stimuli) are regarded as less aesthetic and interesting (Eckblad 1980; 1981). Finally, perception of an aesthetic object is accompanied by a *hedonic tone* of experience; emotions of pleasure and joy.

To sum up, according to Reber and collaborators, the properties of an aesthetic object as a stimulus are *figural goodness (gestalt), figure-ground contrast, stimulus repetition, symmetry and clarity*; whereas the properties of the 'suitable perceiver' of the aesthetic object are *experience with stimuli* (repeated exposures, recognition), *implicit learning*, the *prototypicality* of the object, and the object's *simplicity or complexity*. And both sets of factors, those connected to the object and those connected to the subject, interact during an evaluation of aesthetics. Yet, I would like to add the following two factors about the perceiving subject; that the experience with the object should preferably be multi-sensorial; tactile, auditory, and visual – and the obvious, that the perceiving person's sensory organs should be functioning, both peripherally, and centrally in the nervous system. Regarding the visual perception, both the subject's eyes and the visual cortex should be reasonably 'in order' and functioning. Finally, Plato's demand of a *pleasurable experience* or as we may say *hedonic tone*, is easy to define in modern language. In physiological terms, it is that which produces pleasant, hedonic, experiences in a perceiver; namely whatever activates the cortico-basal ganglia-thalamo-cortical loop and the dopamine pathways in the brain. In a psychological language, one would simply say whatever functions as a reinforcement or reward. In an everyday language, something that makes you feel good.

Perceptual fluency

What all these factors pertaining to the object, and characteristics, or one could even say qualifications, of the perceiving person, add up to, is *perceptual fluency* (Reber *et al.* 2004). That is the ease by which something is perceived. The easier it is for a person 'to grasp' an object, see it, and evaluate it, the more likely it is that that person will regard the object as aesthetic. It is almost self-evident that perceptual fluency must depend on both characteristics of the object and of the perceiving person. But what is new is that as these characteristics are clearly defined, measurable, and testable, what Reber and collaborators have established, is what I will call *objective criteria for subjective evaluations*.

Turning back to my questions whether it is possible to establish whether something is aesthetic or not, and whether that 'something' was made with the intention of creating it aesthetic, I believe that these objective criteria for subjective evaluations can be very useful. I will thus claim that: *De gustibus verum EST disputandum.*

Artefacts put to the aesthetic-test

To choose artefacts as examples to test for their aesthetic properties, their potential for perceptual fluency, is both an easy and a difficult task. There are so many to choose from. And, when I want to explore whether they are 'artful artefacts' it is impossible to avoid the question: Was this or that object made for a function, or was it made purely as art? My personal position is that it is hard to regard works of art as purely and simply 'art'. What we today label as 'art' is decorative, communicative, referential, provocative, expressive, explorative ... etc, and as such, have their particular functions, although not necessarily *practical* functions. To discuss whether products and creations of former times were 'works of art' or 'artful practical objects' is even more difficult, almost futile. For example, Stone Age paintings in caves and icons in medieval churches were both art and functional objects in religious-cultic settings. A pottery teacup and a gold chalice are both artful and useful for drinking. It is often meaningless to try to make a distinction between what is artful versus useful. Many objects are both. But that being said, I still think that an aesthetic quality in objects is 'the little extra' that makes the objects artful and aesthetic, and raises them above the purely functional aspect of their use.

I selected two artefacts as examples for my present analysis. They are not unique specimens, as each is drawn from a class of similar or resembling objects.

Example I. A bifacial pointed stone tool from the Middle Stone Age Still Bay period, 77,000–71,000 BP, Blombos Cave, South Africa

The first example is a bifacial pointed stone tool from the Blombos Cave, South Africa, *c.* 77,000–71,000 years old, from the Still Bay period, Middle Stone Age (Fig. 2.1; Mourre *et al.* 2010). It is *c.* 8 cm long, 2.5 cm wide, and 1 cm thick at its thickest point. Bifacial points like this one were made from flint stones of various colours. Some flints, like this one, were first heated in fire (heat-treated) in order to make them more malleable for knapping and shaping. They also have fine modifications, retouches, on their thin edges. Such points could be shafted to sticks of various lengths and be used as spears to hunt both terrestrial animals and fish in the shallow ponds along the rugged South African coast. As archaeologists we may be astounded by their age and surprised by the effort put into making them. Although we are professional, we are also *subjective* and experiential in our evaluation. We may thus be inclined to regard these points as beautiful, also this selected example. Therefore, I pose the question: how can this artefact be evaluated when tested against the *objective* criteria for subjective evaluations?

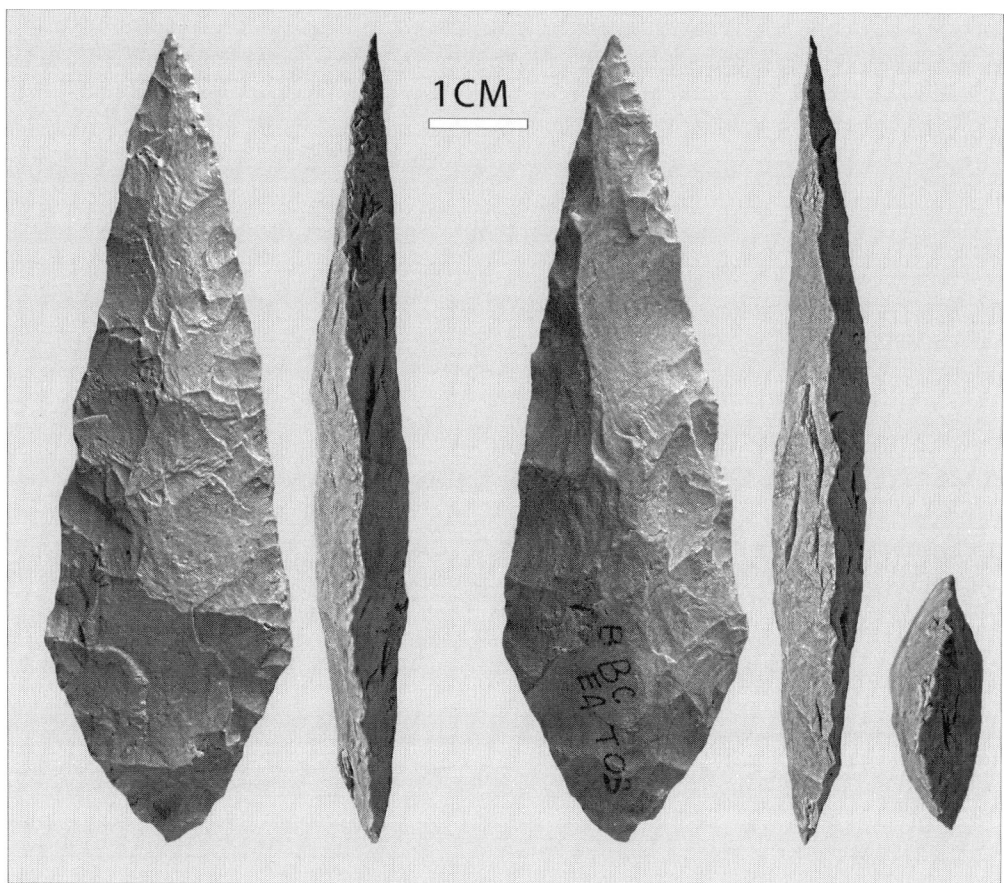

Figure 2.1: A bifacial pointed stone artefact, shown from four sides, Middle Stone Age, from Blombos Cave, South Africa (photo: Magnus Mathisen Haaland)

Does this artefact have *figural goodness* (gestalt)? I think it is reasonable to say that it has a particular form, a shape that makes it stand out as an object. It also has *figure-ground-contrast*, as it stands out as an object against its background. If we consider its original locations, the background against which it would be perceived, may have been sandy ground surfaces, the stone surfaces of caves, and stones of natural shapes. It would clearly stand out against these backgrounds and is distinctly different from other stones and pebbles in nature. It looks 'shaped' and manufactured. Further, it has a shape of *symmetry*, when seen from both its main sides, less when viewing it from its edge, top, or bottom. This artefact is not very complex, but has good *clarity*, an aspect that connects to its figural goodness and symmetry. This artefact is picked out as a singular object, so that the criterion of stimulus repetition may not be relevant. However, in its original contexts, during the time it was used, it would likely be one in a group of similar objects (Fig. 2.2). The hunters and gatherers of 70,000 years ago would likely have had kits with several bifacial stone points of similar forms but varying colours and sizes, depending on their different uses (the size and location of animal or fish to be hunted, for example). So, in its original contexts it may have been one of many and there and then had *stimulus repetition* on a perceptual level, but not when observed alone. I think it reasonable to conclude that this bifacial pointed stone tool passes reasonably well the test regarding the objective criteria for being aesthetic, although with a less than perfect symmetry particularly when observed from its narrow sides, and not displaying stimulus repetition when seen alone instead of as one part of an assemblage of similar artefacts.

When it comes to the perceiving subject, I will try to take the perspective of both an ancient and a modern perceiving person. First the modern. A modern person (except perhaps an archaeologist) would not have had much *experience* with this kind of object, therefore would not have any *implicit learning* connected to it, and would not be able to evaluate the *prototypicality* of it, unless that person had seen other resembling pointed stone tools (Fig. 2.2). Regarding the *simple/complex* criterion, a modern person would probably regard it as simple in form, but with a more complex structure at its surface and edges. These factors: no implicit learning, no knowledge of its prototypicality and 'mixed feelings' about its simplicity versus complexity, could lead a modern subject to regard this artefact as not particularly aesthetic. But, on the other hand, modern subjects are used to and exposed to an enormous variety of objects and forms, so that the very simple outer form, its heterogeneous surface, and the natural material of this artefact might, in itself, lead to a positive evaluation of beauty. Needless to say, its age might also lead a modern viewer to regard it with awe and respect and, therefore, lead to the evaluation that it has high aesthetic quality. Finally, I think that this artefact's objective characteristics – gestalt, symmetry, figure-ground contrast, and clarity – would tend to overrule the wanting characteristics of the modern perceiving subject. Meaning that a modern person would probably enjoy the point's rough aesthetics.

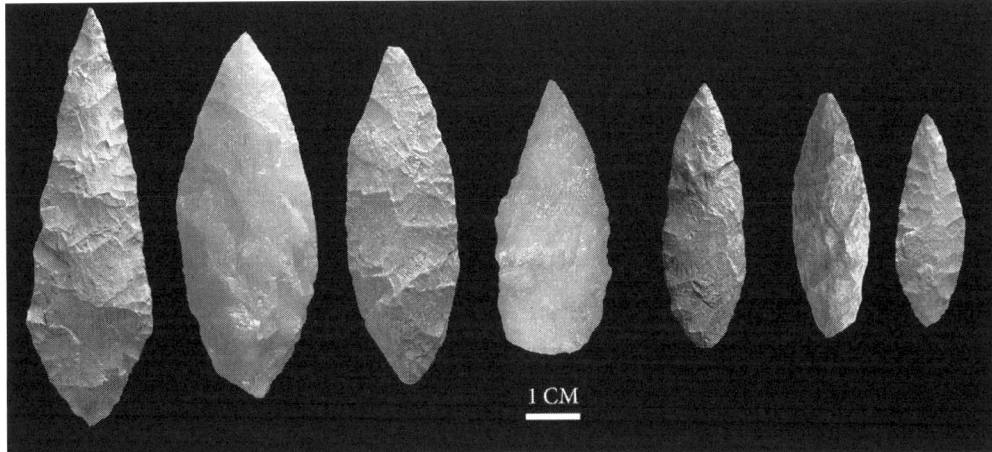

Figure 2.2: Seven bifacial pointed stone artefacts, Middle Stone Age, from Blombos Cave, South Africa (photo: Magnus Mathisen Haaland)

Now I will try to envision the subjects that originally perceived (and used) this bifacial stone artefact. An ancient, contemporary, perceiving person would have a different perceptual approach to it. He or she would most probably have had a lot of *experience* with such artefacts, leading to complex *implicit learning* about it. This person would be fully competent to evaluate the *prototypicality* of it while mentally or factually comparing it to other objects of more-or-less the same class. In the same process, this person would be able to evaluate whether it should be regarded as *simple or complex* by mentally or in reality, comparing it with others of the same class of stone tools. Contemporary, ancient, persons would be likely to process and evaluate this object perceptually with greater ease, and simply experience greater *processing fluency* when looking at and touching it. On these grounds, I think that they would be likely to regard this object as aesthetic. Yet their experience with, and competency regarding, such objects could also have made them more critical in their evaluation. Still, despite the fact that that they may have perceived objects like this one as part of their daily and ordinary world, their tools and things, it is still possible that they would have perceived objects like this as something special and as exceptionally beautiful.

The heat-treatment of the flint before making the tool, may have been simply to improve the malleability of what, originally, may have been a flint of less than perfect quality. However, this treatment also rendered it in an improved state for making a refined finish to it; the delicate and minute retouching (small flaking) on its edges. This was done by a highly controlled pressure-flaking. The process of this manufacture is evident. What is less evident is the motive for doing so. One thinks of early hominids as people who lived hard lives always striving for survival, not as people who had plenty of food and plenty of time to enjoy and explore life and express creativity. However, there is reason to believe that the peoples, anatomically modern *Homo sapiens*, who occupied the South African caves and rock shelters, such as Blombos Cave 77,000–71,000 years ago, actually lived such good lives. They had abundant resources for food and for making implements. Henshilwood claims that these peoples would only need to spend a few hours per day to get enough food (Christopher Henshilwood, pers. comm.). So, they would have had that extra time that was needed to make very elaborate flint artefacts. They had time to make their things beautiful. They had time for making aesthetic artefacts.

Although the bifacial pointed stone tools were very useful, and most of them probably have been used, it is astonishing that they are made more elaborate than necessary for their function. A simpler version would 'work just fine'. What could be the motive for this elaboration? In the analysis above, it is established that the selected specimen of these bifacial pointed stone tools fulfils the criteria for being aesthetic. I think we may conclude that this artefact, as well as the other similar ones, were made *intentionally* to be beautiful. The minute pressure flaking made them both look and feel aesthetic. The emotions connected to making them and, in particular, perceiving them, must have been pleasurable. It is very hard to imagine the opposite. This, and similar incredibly ancient bifacial points were not only useful artefacts but also aesthetic artefacts. I think one might conclude that both modern and ancient perceivers would say they are aesthetic, artful artefacts.

Example II. A brooch from the Migration Period 400–500 CE, Norway

The second example is a relief brooch from the Migration Period, from Kvåle in Sogn, Norway (Figs 2.3 & 2.4). It is made in the so-called 'Animal Art' Style I (Nissen Meyer 1934; Kristoffersen 1995). This style existed during the 5th and 6th centuries in Scandinavia, northern parts of the European continent, and in parts of England (Salin 1935; Hougen 1936). The Migration Period was a period of great political, social, and cultural upheavals and changes, and changes in both populations and elites. The label 'animal art' was given because several animals and birds are found in the patterns. Several objects were made in this style; the most well-known are the large relief brooches and scabbard mountings. Each brooch was unique, as they all were different from others of the same class of brooches. They belonged to a class of objects with styles that were also slightly different in various regions, and the styles changed rapidly over time. To illustrate the differences, another example is shown in Figure 2.5. It is a relief brooch from Dalem, Nord-Trøndelag, Norway (Rygh, 1999, fig. 259).

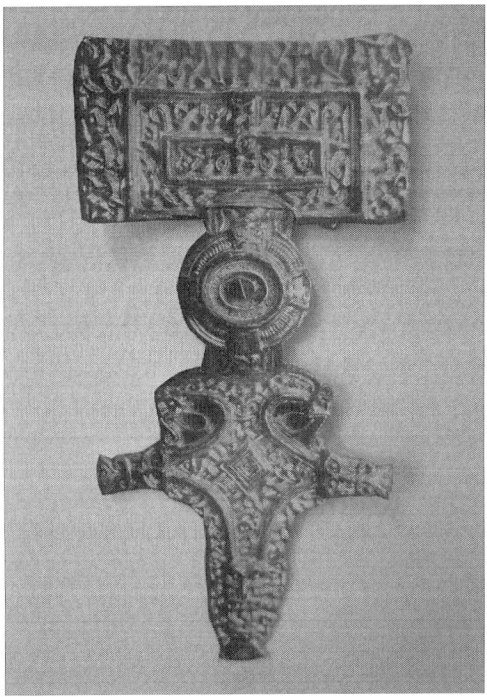

Figure 2.3: A gilded relief brooch, Migration Period, from Kvåle in Sogn, Norway (photo: Torill Christine Lindstrøm)

The brooch selected to be analysed here was cast in iron and gilded. It is *c.* 17.3 cm long and *c.* 8.7 cm wide. The whole brooch is covered with an intricate pattern of figures. The figures are strangely difficult to discover, not only because they are small, but particularly because they are embedded, ambiguous and reversible (Lindstrøm & Kristoffersen 2001). Embedded means that the figures are located within each other, they are ambiguous in that they can be interpreted in various ways, and connected to this, they are reversible in the sense that they, in one moment, can be seen as one kind of object (animal, bird, or person), in the next moment seen as a different object (animal, bird, or person). Parts of one figure function as parts of another as, for instance, where two heads of birds make up a mask or a strange face. The pattern also consists of limbs of birds and animals placed in awkward positions, detached, and twisted, in order to make

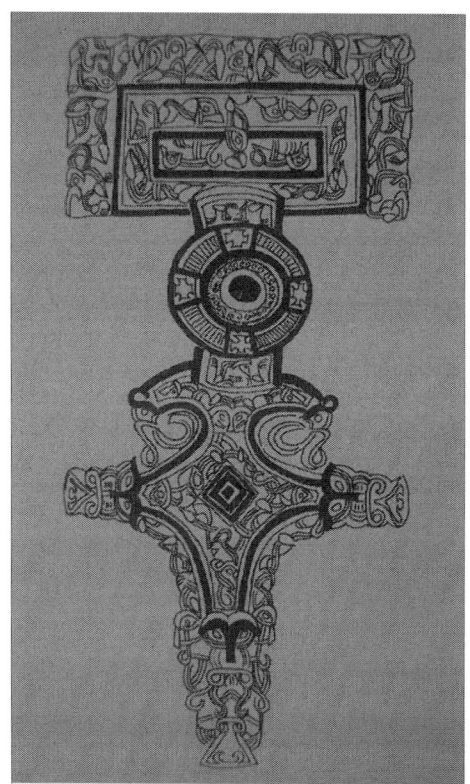

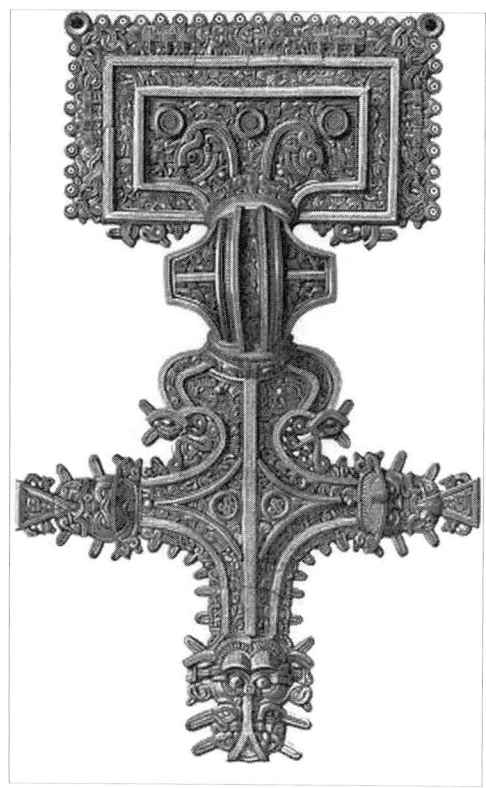

Figure 2.4: The pattern of the gilded relief brooch in Figure 2.3, drawn (drawing: Siv Kristoffersen)

Figure 2.5: A gilded relief brooch, Migration Period, from Dalem, Norway, drawn (drawing: Oluf Rygh)

the patterns integrated, 'swinging' and ornamental. One can spend much time trying to figure out all the figures. New creatures and new combinations of elements emerge. The intricate pattern of the brooch along with the deep and shiny relief, gives the surface 'life' and the figures can be experienced almost as moving.

This brooch was definitely an object of utility, used to fasten a cloak. In addition, it clearly was a symbolic object, indicating social and perhaps even cultic status. The patterns most likely are symbolic and refer to religious rites and myths (Lindstrøm & Kristoffersen 2001). This brooch was a large, shining and glittering, definitely 'catchy' item, most likely meant to impress the viewers, and to give status to the wearer. But is it really aesthetic? Let us put it to the test.

Does this brooch have *figural goodness (gestalt)*? The answer must probably be 'no' as this artefact appears as a composite object with a square plate on top of a more pendant-like shape with a circular part between them. This mixture of forms reduces its gestalt-property. Does it have *figure-ground-contrast*? Yes, it does, despite its poor gestalt-property, because it would stand out against any background as a very conspicuous object. Its gilded surface adds a lot to this. Is there some *stimulus repetition* to be found? Well, that depends on how one defines the stimulus. If the whole brooch is the stimulus, the answer is 'no'. This brooch was carried alone. It was not one among many or one of a pair. However, *within* the patterns of the brooch one might say that there are stimulus repetitions, as some forms and creatures are repeated in the patterning – mostly in a reversed, 'mirrored' manner. That leads to, and connects to, the next demand, that of *symmetry*. This artefact definitely has symmetry, both in its outer form and contours, and within, in the patterning. All the figures are symmetrically shown on both sides of the middle-axis of the brooch – and it is actually this symmetry that, in combination, creates the creatures that can be seen in the middle of the brooch and in the minor details on its edges. Yet, in some small details, the symmetry is subtly broken, possibly to make the patterns even more spellbinding. Finally, whether this brooch has perceptual *clarity* is again, like that of gestalt, a difficult question. Its outline is a mixture of square, round, and curved shapes, and as such does not give an impression of clarity. It definitely has a very intricate patterning which, on a superficial level, may be 'clear' but, on a more accurate level, is 'unclear' for an observer who is unaccustomed to this kind of object and patterning. However, for an observer who is used to this kind of object, it may be clear and 'readable'. So, regarding the criteria for the *object itself*, it may be evaluated as having medium aesthetic qualities. And, in particular, the ambiguous property regarding *clarity* eventually leads us to the aesthetic-evaluation based on the qualifications of the *observer*, the perceiving person.

A modern person, when evaluating the brooch, would most probably be unaccustomed to the embedded, reversible, and ambiguous patterns, and ignorant of the symbols they may convey. Meaning a modern person would lack *experience with the stimuli* of the brooch. A modern person could be confused, bewildered, and perhaps simply uninterested in looking properly into the brooch's patterning. Further, this person would not experience *implicit learning*, of rules that pertain to it, nor be able to evaluate the degree of *prototypicality of the stimuli* of the brooch, unless that person had special competency in the art of the Migration Period. A modern person would probably regard it as complex, both in exterior form and interior pattern. It is therefore likely that a modern person would not experience perceptual fluency or regard the brooch as aesthetic. However, its large size and shining, glittering gold surface might be so impressive that it might overrule the lack of processing fluency. So, it might nevertheless be evaluated as beautiful, but then quite literally on its superficial characteristics: its surface and size.

Contrastingly, a contemporary Migration Period person might experience this brooch, and similar artefacts, as highly aesthetic, and be able to process it, 'read' it, with great fluency. But this on the condition that the evaluating person had had *experience* with this kind of object, opportunities for close-up inspections, and then achieved the *implicit learning* that this experience could bring. Such a person could be able to compare it (in reality or in imagination) with other brooches and thus be able to evaluate it with regard to *prototypicality* and *complexity*. I think that a contemporary person would regard the brooch as prototypical (resembling others) despite its poor gestalt-properties. In particular, a person that was learned and skilled in the symbols of the contemporary mythology, rites and cultic practices would be able to fully evaluate this brooch and its patterns and their symbolic meanings – and far better than any modern person, even better than modern experts on the Migration Period mythology, symbolism and cultic practices. So, I find it reasonable to conclude that a contemporary person most possibly would evaluate this brooch as more aesthetic than would a modern person. And, more than modern persons, contemporary persons would likely be more impressed by its immediate appearance, as they would be less used to seeing shiny, metallic objects and surfaces. So, it is likely that both ancient and modern persons would regard this brooch as a beautiful, artful artefact, although for somewhat different reasons.

Why aesthetics? Why artful artefacts?

Aeons of time separate the two objects presented and analysed here. The Middle Stone Age Still Bay bifacial pointed stone tool from Blombos Cave in South Africa and the Migration Period brooch from Kvåle in Sogn in Norway are from distinctly different times and cultures. Nevertheless, on as objective grounds as it is possible to establish, they are both likely to be evaluated as aesthetic, both by modern persons and by the contemporary persons who produced and used them or at least perceived them.

Regarding the beauty of the Middle Stone Age bifacial pointed stone tool, the same conclusion, in this case concerning Acheulean hand-axes made by *Homo erectus*, was presented by Wynn and Berlant (2019, 278):

> Most stone-tools were straightforward mechanical devices that hominids made and used to perform work. But some appear to have been overdetermined; hominids invested more time and effort in their manufacture than was necessary for their mechanical function.

It is this 'added value' that has the potential to inform us about the makers' and users' aesthetic sensibilities. So apparently, these early humans, from *Homo erectus* to early *Homo sapiens* at Blombos, had both the ability to enjoy aesthetic properties of objects and the ability, time, and willingness to produce over-done, over-determined, beautiful, aesthetic, artful artefacts.

The Migration Period brooch was also over-done and over-determined. A far simpler needle, hook, or fibula could hold a cloak together. For that function, both its size and decorations were exaggerated. A lot of effort and resources were invested in making it beautiful and fascinating. Contrasting the Stone Age artefact, where the maker's intention regarding aesthetics is implicit and must be inferred, the aesthetic and artistic intention of the brooch's maker is explicitly evident.

Living in our time, when functionality and automated, economical procedures are overriding ideals in the production of most of our objects, implements, and gadgets, intentions of making things beautiful call for an explanation, in particular when that would require a lot of extra effort and personal skill and take place in cultures that had much simpler technologies than we have. What could their motivations be? Why aesthetics? Why making artful artefacts? Was there a meaning connected to them? Were there symbolic or semiotic functions? If so, what could they have been?

Here, we reach the limits, and the limitations, of studying ancient artefacts solely as 'material culture'. Here we enter into the realm of the 'immaterial culture' – and the realm of psychology.

Motives for making aesthetic artefacts

Motives are intrinsic or extrinsic. Intrinsic motivation means that a behaviour or production is rewarding in itself. Simply doing it, performing it, or creating it, gives rise to fascination, absorption, and hedonic emotions, sometimes even euphoria. Extrinsic motivation means that a behaviour or production is rewarded from 'outside'; it results in status, admiration, or more concretely 'pays off'; it is monetary paid for, can be used in exchange for various goods, and can attract potential partners. But clearly, a person's motivation can be a mixture of intrinsic and extrinsic motivating forces. The act of creating, where one's skill, creativity, phantasy and imagination can unfold, is intrinsically motivated – yet, thoughts about external rewards, from admiration to various forms of payment, may be present, and those are externally motivated.

Production of art and artful artefacts is pervaded with both intrinsic and extrinsic motivations. The act of creating can be intrinsically enjoyable while also being motivated by the expectancy of an external reward. Admittedly, abnormal mental states have also been the backdrop of creativity and works of art. When making the bifacial pointed stone tool, its maker may have entered a state of compulsive perseveration where the flaking was hard to stop, resulting in the fine cuts on the object's edges. Yet, since several resembling specimens are found, this explanation is unlikely.

Regardless of its maker's motives, intentions and creativity, the bifacial point must have been a product in, and of, a tradition, with possibly also collective intentions and motivations at work. The intrinsic pleasure of making and looking at something beautiful could be one factor. Another is the possible admiration and status given to the persons and the culture that could produce such artful, aesthetic artefacts. As said above, admired objects might be used as a 'currency', as trade-products in exchanges of goods, and as payment for partners. But one could also muse over more subtle functions, symbolic functions. These objects could have been individualised status-objects or collective signature-objects of this particular culture, cultural semiotic signs; and they might even have been used, or *also* been used, as paraphernalia during cultic rituals.

The brooch was undiscussably meant to be big, beautiful, and impressive. Visibility must have been a motivation and intention. It was made by a skilled professional smith but commissioned, owned and worn by a person of status (Lindstrøm & Kristoffersen 2001). This might have been a person belonging to a high-ranking family or clan, perhaps even a prestigious person with cultic functions and performances, a chieftain, priest or priestess. The patterning shows animals and birds of power: eagles, wolves, horses, dragons, bears, as well as bearded persons. All are symbols of power, possibly also religious symbols. The intricate designs with embedded, ambiguous and reversible figures are interpreted as referring to cultic practises of transformation. This implies that a human person (a priest, priestess, shaman, warrior) could transform into an animal or bird (shapeshifting) and get access to their superior abilities and powers (Lindstrøm & Kristoffersen 2001; Kristoffersen 2010; Lindstrøm 2012). It is therefore possible that only persons who had knowledge of these symbols and practices could fully appreciate the semiotic aesthetics in patterns on a brooch, as the one that I have analysed here, and be a qualified 'suitable perceiver'. For that person, the perception was not only a registration but also an experience. An aesthetic experience, an experience of art – and perhaps even a revelation.

Unlike the bifacial points, this brooch and others like it were unlikely to be objects of trade, but might be valuable inherited objects, transferring both value and status to the next generation. And, perhaps needless to say, like the owner of beautiful bifacial points in the Stone Age, an owner of an animal art brooch in the Migration Period might attract attention from others, also from potential partners. Despite the time and distance separating the bifacial point and the brooch, they are both aesthetic objects, and even some of the motives for making them aesthetic, and some of their symbolic functions, may have been similar.

The immaterial cultural aspects, regarding both the point and the brooch, are invisible to us. The archaeologists' creative imagination and ephemeral aesthetic experiences in encounters with artful artefacts are necessary for attempts to decode such objects' subtle invisible aspects; as well as for trying to envision and understand the makers' and users' minds.

Aesthetics encounter art

'Aesthetics and art are not co-extensive' (Wynn & Berlant 2019) – and the concept of 'art' is problematic in itself. It is very unclear when people started to regard 'art' as a class of phenomena in itself, in contrast to 'decoration' and 'beautiful thing'. And indeed, not all 'art' is beautiful. Some art may be hideous in its display of cruelty, death and desperation, but still have aesthetic properties. Other art may simply be ugly, disgusting and repulsive, and was intended to be perceived as such. The intentions and motives of the makers are varied. In particular, during the last 100 years, ideas of what counts as 'art' have been completely transformed.

Contrastingly, when we talk about aesthetics, we think about beauty. That which is 'pleasing to the eye', that which is perceived with processual fluency, and that which gives the perceiver 'a pleasurable experience'. Art theory and aesthetic theory are only partially compatible (Hepburn 2009). The feeling of aesthetics, of beauty, is an emotionally salient response, important in itself. Apparently, now-living non-human primates do not, and primeval hominids did not have any recognition of, or sense for, aesthetic properties of their tools. A sensitivity to aesthetics seems to have evolved along with higher cognitive functions, and in particular in connection with complex intra-group social organisations and inter-group interactions (Wynn & Berlant 2019). A sense for aesthetics and a willingness to start an aesthetic process of production also connect to cognitive abilities of abstraction. And, needless to say, abstraction is the very essence in all representation – and thus in art as representative objects or performances.

Representation or revelation?

A central theoretical position regarding art is that it is representative (Hepburn 2009). Although this concept is philosophically problematic (Wollheim 1980; 1987), it means either that works of art show, or describe in some way, natural or cultural phenomena; or that objects of art refer to something other than themselves. In archaeology, 'representation' is also frequently used to discuss artefacts that copy something or refer to something. Yet cultures differ in their representational modes and codes, in their iconographic programmes, and in their representational practices (Gibson 1979), implying that one encounters great complications in interpreting. In addition, as Ingold wrote about depictions made by hunter-gatherers, depictions may not be 'representations' at all, not something referring to something else but revelations in themselves, so that in order to fully understand, it is necessary 'to penetrate beneath the surface of things so as to reach deeper levels of knowledge and understanding' (Ingold 2011, 130). Could this insight also apply to the bifacial point and the brooch? Possibly. The bifacial point does not 'represent' the skill and competency of its maker and the culture he or she belonged to, it *reveals* it. Likewise, although not produced within a hunter-gatherer culture, the brooch *reveals* the skills of the smith who produced it, but it also *reveals*, through its semiotics, important ideological concepts and ritual practices of the culture that produced it. As the transformations actually take place within the pattern, the complex pattern is a *revelation* in itself.

Again, I claim, we see here the limitations of defining and practising archaeology as the science of 'material culture' only. In most artefacts we encounter immaterial culture, perhaps particularly in art and aesthetic artefacts. Every culture has 'what goes without saying', and that which is beyond saying is hard to express in words. The ineffable and intangible 'meanings' in a cognitive sense, and in a collective cultural sense are values, visions, belief systems, and codes. This is the realm of symbolism, semiotics and art.

Archaeology reaches a threshold where perception and understanding of artefacts, particularly artful artefacts, can only be complete and receptively consummated within an experiential, intuitive, open and receptive mode of perception. Only through an endeavour to reach a holistic mode of imaginative mental processing can the invisible cultural traits, the *intangible* elements of a culture, be sought and understood. Of course, the basic archaeological elements have to be in place: dating and classification, and the historical, ideological, and social parameters must be recognised. But a full, or at least a fuller, comprehension can only be reached within the imagination of the archaeologist. In an explorative, artistic mode, but still within a scientific sober mode, we should try to reach out for the psychological factors, the cognitions, sentiments, and emotions of the persons behind the objects – those who produced and used the artful, aesthetic artefacts.

Acknowledgments

This work was supported by the Research Council of Norway through its Centres of Excellence funding scheme to the SFF Centre for Early Sapiens Behaviour (SapienCE), Project 262618.

References

Arnheim, R. 1974. *Art and Visual Perception: The new version.* Berkeley CA: University of California Press.

Berlyne, D.E. 1971. *Aesthetics and Psychobiology.* New York: Appleton-Century-Crofts.

Berlyne, D.E. 1974. *Studies in the New Experimental Aesthetics. Steps Toward an Objective Psychology of Aesthetic Appreciation.* Washington DC: Hemisphere.

Birkhoff, G. 1933. *Aesthetic Measure.* Cambridge MA: Harvard University Press.

Eckblad, G. 1980. The curvex: Simple order structure revealed in ratings of complexity, interestingness, and pleasantness. *Scandinavian Journal of Psychology* 21(1), 1–16.

Eckblad, G. 1981. Assimilation resistance and affective response in problem solving. *Scandinavian Journal of Psychology* 22(1), 1–16.

Ekehammar, B. 1974. Interactionism in psychology from a historical perspective. *Psychological Bulletin* 81(12), 1026–1048.

Eysenck, H.J. 1941. The empirical determination of an empirical formula. *Psychological Review* 48, 83–92.

Fechner, G.T. 1876. *Vorschule der Ästhetik.* Leipzig: Breitkopf & Härtel.

Garner, W.R. 1974. *The Processing of Information Structure.* Potomac: Lawrence Erlbaum Associates.

Gibson, J.J. 1979. *The Ecological Approach to Visual Perception.* Boston MA: Houghton Mifflin.

Gombrich, E.H. 1984. *A Sense of Order* (2nd edn). London: Phaidon.

Gombrich, E.H. 2014. *The Story of Art* (16th edn revised). London: Phaidon.

Hepburn, R.W. 2009. Theories of art. In D.S. Cooper (ed.), *A Companion to Aesthetics.* Oxford: Blackwell, 565–569.

Hougen, B. 1936. *The Migration Style of Ornaments in Norway.* Oslo: Universitetets Oldsaksamling.

Humphrey, D. 1997. Preferences in symmetries and symmetries in drawings: Asymmetries between ages and sexes. *Empirical Studies of the Arts* 15, 41–60.

Ingold, T. 2011. *The Perception of the Environment. Essays on Livelyhood, Dwelling and Skill.* London: Routledge.

Kristoffersen, S. 1995. Transformation in migration period animal art. *Norwegian Archaeological Review* 28, 1–17.

Kristoffersen, S. 2010. Half beats – half man. *World Archaeology* 42, 261–272.

Lindstrøm, T.C. 2012. 'I am the walrus': Animal identities and merging with animals – exceptional experiences? *Norwegian Archaeological Review* 45, 151–176.

Lindstrøm, T.C., & Kristoffersen, S. 2001. 'Figure it out!' Psychological perspectives on perception of migration period animal art. *Norwegian Archaeological Review* 34(2), 65–84.

Mourre, V., Villa, P. & Henshilwood, C.S. 2010. Early use of pressure flaking on lithic artifacts at Blombos Cave, South Africa. *Science*, New Series 330, 659–662.

Nissen Meyer, E. 1934. *Relieffspenner i Norden*. Bergen: Bergens Museums Årbok 1934/Historisk-antikvarisk rekke 4.

Reber, R. 2002. Reasons for the preference of symmetry. *Behavioral and Brain Sciences* 25, 415–416.

Reber, R. & Schwarz, N. 1999. Effects of perceptual fluency on judgments of truth. *Consciousness and Cognition* 8, 338–342.

Reber, R. & Schwarz, N. 2001. The hot fringes of consciousness: Perceptual fluency and affect. *Consciousness and Emotion* 2, 223–231.

Reber, R., Schwarz, N. & Winkielman, P. 2004. Processing fluency and aesthetic pleasure: Is beauty in the perceiver's processing experience? *Personality and Social Psychology Review* 8(4), 364–382.

Reber, R., Winkielman, P. & Schwarz, N. 1998. Effects of perceptual fluency on affective judgments. *Psychological Science* 9, 45–48.

Reber, R., Wurtz, P. & Zimmermann, T.D. 2004. Exploring 'fringe' consciousness: The subjective experience of perceptual fluency and its objective bases. *Consciousness and Cognition* 13, 47–60.

Rygh, O. 1999. *Norske Oldsager*. Trondheim: Tapir.

Salin, B. 1935. *Die altgermanischhe Thierornamentik. Typologische Studie über germanische Metallgegenstände aus dem IV. bis IX. Jahrhundert, nebst einer Studie über irische Ornamentik*. Stockholm: Wahlström & Widstrand.

Solso, R.L. 1997. *Cognition and the Visual Arts*. Cambridge MA: MIT Press.

Tatarkiewicz, W. 1970. *History of Aesthetics. Vol 1, Ancient Aesthetics*. The Hague: Mouton.

Wollheim. R. 1980. *Art and its Objects*. (2nd edn) Cambridge: Cambridge University Press.

Wollheim, R. 1987. *Painting as an Art*. London: Thames & Hudson.

Wynn, T. & Berlant, T. 2019. The handaxe aesthetic. In K.A. Overmann & F.L. Coolidge (eds), *Squeezing Minds from Stones*. New York: Oxford University Press.

Chapter 3

The importance of the anthropological approach in archaeology: the example of prehistoric acoustic studies

Iegor Reznikoff

In contradistinction to pure sciences (e.g. mathematics or geology), in applied sciences, studies and results are not to be stated blindly, *per se*, as pure results (mathematical or geological) of the science they are based on, ignoring the proper domain where this science (maths or geology) is to be applied. The study of the proper domain is grounded first on an anthropological approach; the initial approach, even before the subject becomes an 'applied science', comes from an experience essentially based on direct perception. For instance, the importance of eyes and vision in archaeological studies is obvious, as for the discovery of flint tools or of a hidden statue. When in painted caves or on painted rocks one discovers paintings or engravings, it is first with eyes, and when a painting is said to represent a horse or a bison, it is also, clearly, with the help of eyes; it would be crazy to try to discover this with machines. Of course, various devices can be used, later on, for chemical analysis of colours or for dating, but the first approach is of anthropological character. Curiously enough, it is not the case in archaeoacoustics. After the discovery, in 1983 in prehistoric painted caves, of the relationship between locations of paintings and acoustics of these locations (to state it simply: the more a location sounds, the more pictures in it) (Reznikoff 1987a; 1987b; Reznikoff & Dauvois 1988) and, in 1987 of the relationship, in open air, between rock paintings and echoes (Reznikoff 1995; see also Reznikoff 2014a), these discoveries opened a new era, namely of acoustics applied to prehistory. Many studies have been carried out recently (for instance, R. Raino *et al.* 2017 and M. Diaz-Andreu *et al.* 2019, in open air, and, in painted caves, R. Till 2017 compared with Reznikoff 2014b). However, these last studies, contrary to the seminal one (Reznikoff 1987a) were undertaken only by using machines without any anthropological approach. This limits seriously the scope of investigations and depth of results. The question remains as to how it was in prehistoric times since ancient tribes, of course, had no machinery for making sounds and certainly not for recording. Moreover, keeping in mind the point of view of the present volume, *Art in the Archaeological Imagination*, the question of a possible artistic approach with the machinery used obviously has no meaning. Contrariwise, in order to understand how an anthropological approach may have some artistic aspect, let us give some examples.

Recently (June 2019), with a team under the direction of Riitta Raino (University of Helsinki), we studied acoustics around a painted rock on lake Humaljärvi, near Lapeenranta, in Finland. (Fig. 3.1) The study was done using voice and acute attentive hearing (for more explanation, see below) and, as a complement, a sophisticated recording device for later analysis of reflected sounds. The acoustic quality of the surroundings, measured mostly by the number and quality of echoes, was good and clearly in relationship with the painted rock. The experience was done at various distances from the rock (Fig. 3.2). At a distance of about 50 m the rock answers exactly the syllables one produces while facing it; so that one has the impression that the rock speaks. This corresponds to the widespread belief in many cultures that a Spirit dwells inside the rock. You sing O, it answers O; A, it answers A, and so on; A E O, it answers A E O; a very strange impression. Now, if the acoustic study were to be done with artificial devices (pistols, air balloons or continuous sets of frequencies emitted by loudspeakers) the rock would have answered with a corresponding noise, that is

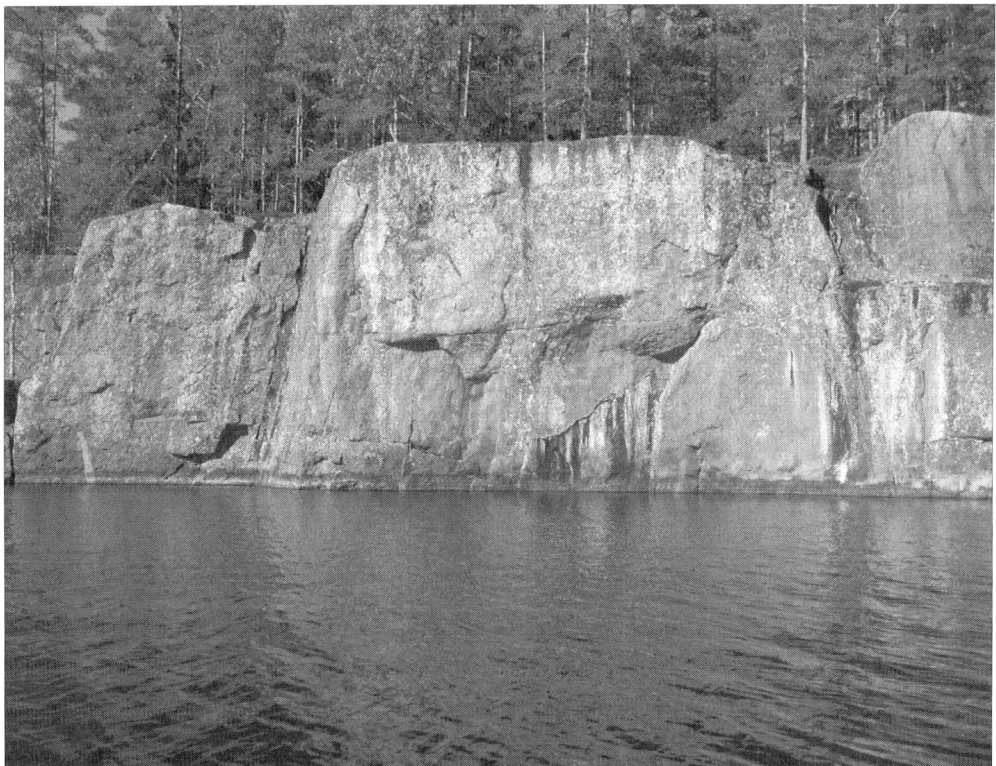

Figure 3.1: The Speaking Painted Rock (photo: Julia Shpinitskaya)

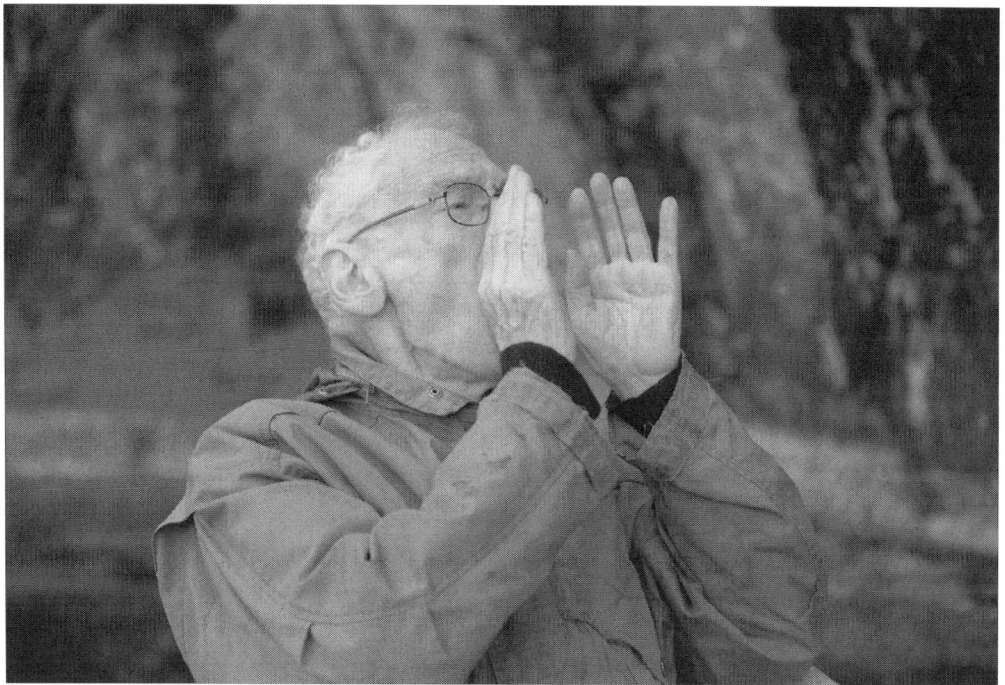

Figure 3.2: Making sounds for echoes (photo: Julia Shpinitskaya)

all. You cannot discover that the rock speaks unless you speak to it; one would not even think of this if working only with artificial devices. The anthropological approach is necessary. Of course, the property of 'speaking' or answering of the rock can be shown by the harmonic quality of the reflected sound but this needs a careful and long analysis of the recorded sounds and is impossible to do on the spot. While with voice and hearing one can immediately try to improve the answer of the rock by varying the precise distance, direction, and intensity of the voice. Moreover, echoes are very sensitive to external conditions such as wind, rain, surface of water, which may change very quickly. The rock we have studied, when there was no wind and the lake was still, was indeed a remarkable speaker (Fig. 3.3); in our approach we could have tried to discover what short sentence and which melody could be the best for the answer, this clearly would be an artistic approach. Actually, we did not try; but we did elsewhere (see below).

We must recognise that recently, with the progress of technology for emission and recording, some advantages have appeared of studies with specific devices in open air. The devices for emission with a sufficient intensity repeated as many times as necessary, and for recording with the possibility of determining the directions of returning echoes, have some advantages on the human possibilities. Such devices were used, for instance, by Margarita Diaz-Andreu *et al.* (2019) in a recent study. Repeating some sounds again and again, with a good intensity (around 90 dB) can be exhausting

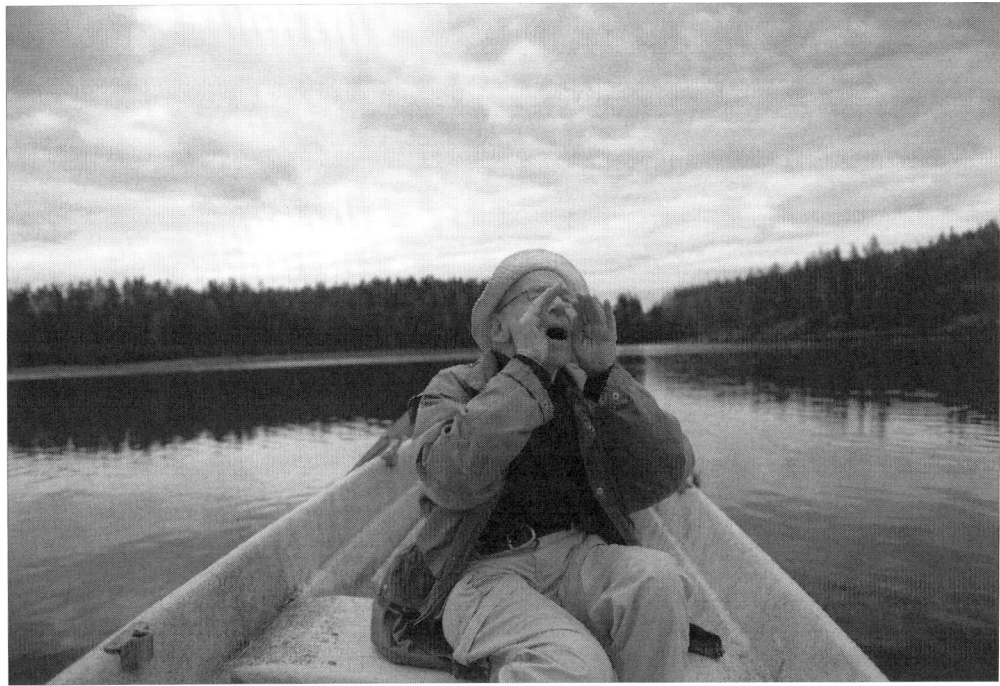

Figure 3.3: Making sounds in the middle of the lake (photo: Julia Shpinitskaya)

for the voice unless you have as many voice producers as people needed to carry and use successfully all elements of the high-tech devices; but actually, it is easier to find such gifted technicians than it is to find persons gifted to produce some vocal sounds and short melodies. However, when using only devices, the crucial question appears to be whether the same results could be achieved anthropologically by using human voices or horns, which was the only possibility in ancient times. Moreover, the human voice has the advantage of possibly singing echoing melodies and permits the choosing of what melodies are best suited for a given space. With the complexity of the change of echoes in time, the impression can be miraculous; for instance when you hear four, five, or more echoes. Or if you sing on the vowel O a melody, say on *c g f a g* (other notes are above *c*), preferably in just intonation, and the echo answers twice, possibly overlapping the same melody, and after a short while answers a third time with a shorter melody, say, *a g a g*, the impression created by sounds and space is pure and deep magic. Indeed, the echoes of echoes have progressively increased the purity of intervals in a decreasing intensity, and pure soft intervals produce a subtle, persistent and beneficial effect on deep levels of consciousness.[1] Of course, such marvellous impressions are important to try to feel how deeply prehistoric tribes honoured painted rocks with rituals and sounds; in such a feeling one may, perchance, make new discoveries.

In painted caves the anthropological approach is actually much more needed because underground closed spaces are usually quite intricate. In open spaces one has often to climb up or slide down on rocks at different levels but, in a cave, the space, sometimes very large (e.g., the main Hall in Isturitz, Pays Basque, France; *Salon Noir* in the Niaux cave, Ariège), can reduce to a narrow tunnel where one has to crawl carefully (tunnel with red dots in the Portel cave, Ariège, or in the Oxocelhaya cave, Pays Basque) and there are often also hidden niches and recesses (Fig. 3.4). All these peculiar locations are generally not accessible to machinery, except your own head, voice and ears! I was once in the cave of Arcy-sur-Cure (Burgundy; Fig. 3.5) with David Lubman, a renowned specialist in acoustics (Lubman 2008), and asked him how he would study the relationship between acoustics and locations of pictures. He answered that he would do as he always does: clapping hands and analysing the recorded return of the sound. Clapping hands produces a 'white noise', i.e. an almost continuous set of frequencies. The resonant frequencies of the space can be discovered only afterwards, as Lubman said, in the analysis of what has been recorded; but this needs special devices and time. I told him that this must be done at different places in the cave, in tunnels, niches, etc, as mentioned above. David Lubman recognised that the usual machinery would be too hard to manage with and the progression would be deaf and blind since, with the machinery, one needs a long analysis of the

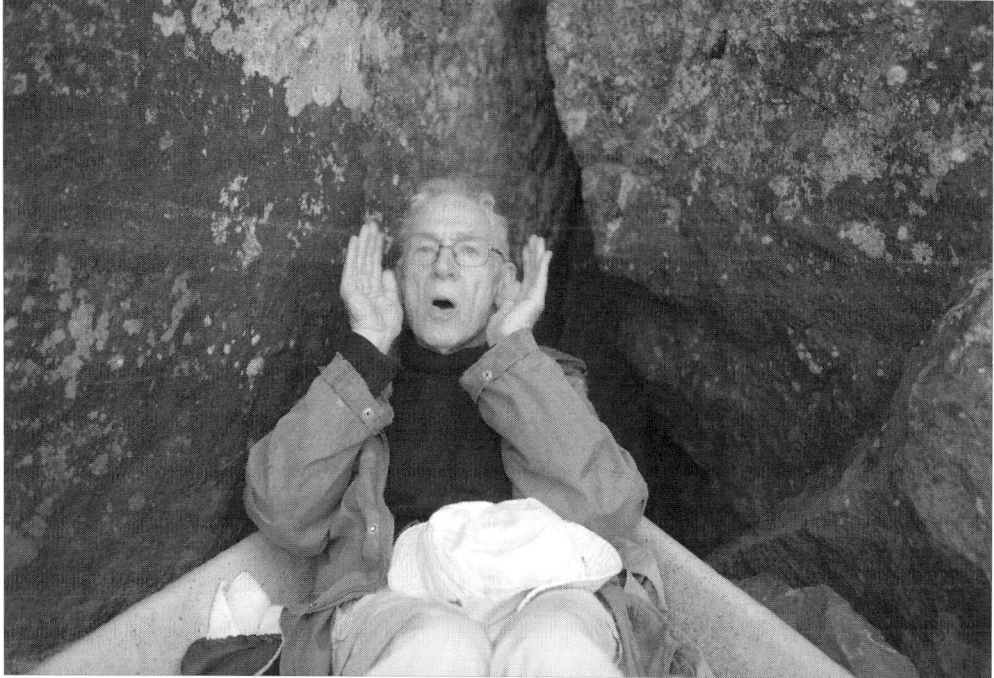

Figure 3.4: Making sounds in a recess of the Speaking Rock (photo: Julia Shpinitskaya)

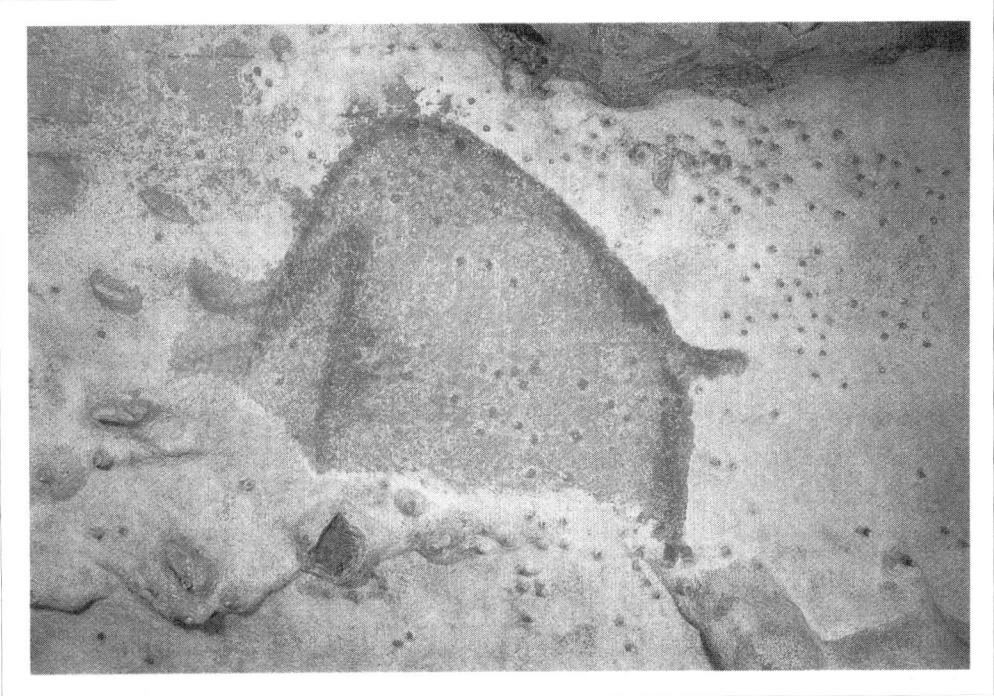

Figure 3.5: Arcy-sur-Cure Cave, 'Mammouth Diamanté' in the most resonant part of the cave (collection La Varende, photo M. Girard)

results in order to discover the resonant frequencies; moreover, in some parts of caves, recording devices have no access.

Certainly, such analysis cannot be done immediately on the spot while clapping hands, shooting with a gun or using inflatable balloons, contrary to the live production of sounds with the voice, since our phonation-hearing ability is particularly subtle and fast. When producing vocal sounds, a simple *mm* at different pitches, the resonant frequencies are immediately heard and felt in the body because these frequencies are strongly amplified by resonance. Recall that our human species is a speaking one. This needs a particularly fine capacity of hearing overtones of different vowels and consonants; moreover ancient tribes, hunting with bow or spears, needed a remarkably fine perception for moving silently and hearing very gentle sounds, or imitating sounds of various birds and animals. This ability was, and is, also necessary to progress in completely dark caves. Ancient tribes had only small oil lamps producing a very limited area of clarity, or torches which cannot be used in niches, narrow tunnels, etc; so that the only way to proceed in this darkness is by making sounds: rather powerful O, O, OHO, and listening to the answer. If there is no answer in one direction, it means there is no space in that direction; if there is an answer with, say, three echoes, this means that there is a large space to progress into and, if the answer comes from underneath, it possibly reveals a hole or an abyss, in which case it is better to withdraw. This, what we call *echolocation*, is a functional use of sounds and if it cannot be qualified as 'artistic', it is certainly anthropological. Using artificial devices it would be practically impossible because you have to react immediately to what you hear, which is very easy by echolocation. Remarkable results can be obtained by (e.g. blind) people trained in echolocation.

A careful attentive progression in the cave can be very rewarding. In Le Portel cave (Ariège), a first picture, an owl, when studied carefully in its acoustic environment, was discovered to be in echo-relation with further images (Reznikoff & Dauvois 1988). The relative importance of the discovery, by bringing attention to the location of the owl, yielded a new discovery; there was another picture hidden behind the owl. This possible echo-relationship between distant pictures or panels, which we have remarked in other caves, is typically something you hear and discover on the spot at the location of a picture while hearing the echo and evaluating approximately from where this echo returns, and finally discovering that it comes from a wall with other pictures. This again would be almost impossible to discover with machines since you need first to record the sounds and then, after analysing them elsewhere, to return possibly several times into the cave, trying to locate the echo, if there really is one. Beforehand you are, as we said, deaf and blind if you are not proceeding with sounds and simultaneous hearing. Moreover, in many caves, the time allowed for research is limited and to go and return several times is excluded.

A fourth example is Le Portel, Oxocelhaya. Here, while progressing in a narrow tunnel, hardly crawling, and trying by humming to find the most resonant locations in the tunnel, you suddenly reach a point with a very impressive maximum of resonance where the whole tunnel sounds. You look with a torch at the nearest wall and discover one or more red dots which happen to coincide with the location of the maximum of resonance (concerning resonance in Palaeolithic times, see Reznikoff 2006). This is very remarkable and impressive but could not be realised and discovered without our own human instant perception. In very resonant parts of a cave and particularly in niches or tunnels as just described, the perception involves the feeling of sounds and vibrations in the whole body. To experience this or to get a strong and unexpected sonorous answer in front of a picture of an animal is an unforgettable experience. It certainly deepens the understanding of the prehistoric universe.

Our last example, probably the most remarkable from the point of view of an artistic creation of sounds, is when imitating growls and roars of animals in a sonorous niche or small alcove. Making simple growls or neighing like a horse on a main low pitch of the resonance in the niche produces a terrific effect, giving the impression that a herd of bison, lions, or wild horses, are growling, fighting or running about as the sound is amplified by the resonance of the part of the cave where the niche is situated.[2] In relationship with pictures of corresponding animals (Niaux, Arcy), the effect – called *bison effect* – is really breathtaking (Reznikoff 2012). Let us recall that imitating cries of animals is characteristic of ancient shaman's incantations in order to reach deep levels of consciousness.

It may seem that the ways of producing the sounds mentioned in this study are too special and difficult to master; but, first, the pitch of a resonance is easy to find because this pitch is immediately amplified by resonance, and, second, it is not a voice for opera singing that is needed but a simple rather low voice (*c.* 90–230 Hz), mastering some elementary tunes, preferably in just intonation, reaching an intensity of 90 dB on high notes. Such a voice is needed for echolocation in order to explore the almost completely dark cave, as we have explained above, but it does not mean that female voices were excluded when, in previously discovered resonant parts of the cave, some rituals with paintings, sounds and probably dances were performed (Reznikoff 2014a). Of course, a fine attentive trained hearing is also required but can easily be acquired. The artistic ability is to be discovered on the field as we have seen in examples above; the performance is in no way artistic *per se*, the aim is obviously different and purely anthropo-archaeological in order to discover and understand better the magic underground universe. We must beware not to put modern fancy artistic tricks, inspired by excited imaginations, on prehistoric studies, but just look at the exceptionally remarkable paintings or engravings and listen to their answer to some vocal sounds, just OO or MM, sung gently but strongly enough to get an answer. There is nothing to add, except profound respect and a wish for a deep and sometimes subtle understanding of the remarkable, artistic, ritual and say, liturgical representation of a visible, audible and invisible world of life, colours, pictures, animals, sounds, humans perhaps, and earth, all reflected by the majesty of rocks.

Notes

1. See http://www.musicandmeaning.net/issues/showArticle.php?artID=3.2
2. Listen to *Le Chant des Grottes Préhistoriques à Peintures*, CD to appear (2020). For the exceptional beauty of a Romanesque resonance, listen to *Le Chant du Mont St-Michel*, CD (SM – ADF Bayard 2001).

References

Diaz-Andreu, M. & Mattoli, T. 2019. Rock art, music and acoustics: A world overview. In B. David & I.J. McNiven (eds) *The Oxford Handbook of the Archaeology and Anthropology of Rock Art*. Oxford: Oxford University Press, 503-528.

Lubman, D. 2008. Convolution-scattering model for staircase echoes at the temple of Kukulkan. *Journal of the Acoustical Society of America* 123(5), 3604.

Raino, R., Lahelma, A., Äikäs, T., Lassfolk, K. & Okkonen, J. 2017. Acoustic measurements and digital image processing suggest a link between sound rituals and sacred sites in northern Finland. *Journal of Archaeological Method and Theory* 25(1), 453–474.

Reznikoff, I. 1987a. Sur la dimension sonore des grottes à peintures du paléolithique I. *Comptes rendus de l'Académie des Sciences Paris* 304 II(3), 153–156.

Reznikoff, I. 1987b. Sur la dimension sonore des grottes à peintures du paléolithique. *Comptes rendus de l'Académie des Sciences Paris* 305, 307–310.

Reznikoff, I. 1995. On the sound dimension of painted caves and rocks. In E.Tarasti (ed.) *Musical Significance, Symposium de Helsinki 1988*. The Hague: édition Mouton de Gruyter, 541–557.

Reznikoff, I. 2006. The evidence of the use of sound resonance from Palaeolithic to medieval times. Acoustics, space and intentionality. Identifying intention in the ancient use of acoustic space and structure. In G. Lawson & C. Scarre (eds) *Archaeoacoustics*. Cambridge: McDonald Institute for Archaeological Research Monograph, 77–84.

Reznikoff, I. 2012. L'existence de signes sonores et leurs significations dans les grottes paléolithiques. In J. Clottes (ed.) *L'art pléistocène dans le monde/Pleistocene art of the world/Arte pleistoceno en el mundo. Préhistoire, Art et Sociétés. Bulletin de la Société Préhistorique Ariège-Pyrénées* 65–66, CD: 1741–1747.

Reznikoff, I. 2014a. On the sound related to painted caves and rocks. In J. Ikäheimo, A-K. Salmi & T. Äikäs (eds) *Sounds Like Theory*. Saarijarvi: Monographs of the Archaeological Society of Finland 2, 101–109.

Reznikoff, 2014b. The Hal Saflieni Hypogeum: A link between Palaeolithic painted caves and Romanesque chapels? In L. Eneix (ed.) *Archaeoacoustics*. Myakka city FL: OTS Foundation, 45–50.

Reznikoff, I. & Dauvois, M. 1988. La dimension sonore des grottes ornées. *Bulletin de la Société Préhistorique Française* 8(85), 238–246.

Till, R. 2017. An archaeoacoustic study of the Hal Saflieni Hypogeum on Malta. *Antiquity* 91(355), 74–89.

Chapter 4

Replicating the prehistoric artisan's mindset

Jacqui Wood

It is generally assumed that the techniques and skills of artisans of all mediums pass down to their apprentices. So when one tries to replicate the artwork that formed so many integral parts of daily life in prehistory it is not so easy as there are no written records. If you are going to replicate something made say from the Bronze Age you do not even have the observations of the classical historians to guide you, as is the case in the Iron Age. Therefore, a different approach is needed to rediscover their skill set, having only the artefacts and in the case of house structures the site plans to work on. Over the last 27 years I have developed a method that has proved successful in unlocking the mysteries of some of the artisan techniques of prehistory.

This chapter will follow the format of an Artists Diary or Journal taking the reader through each process I followed and my discoveries on the way. I am in a way going to impart my skill set as I might do to my apprentice, so he or she can go on to discover those lost techniques for themselves. I will start this process by looking at my research into the manufacture of Bronze Age ceramics in Cornwall in the south-west of England. I have always believed that the inherent skills and personalities of prehistoric people are very much the same as they are today. We have lost most of their practical skills because we don't require them anymore. I believe that if civilisation as we know it today was destroyed it would take less than a generation for us to re-acquire the same skill set as the average prehistorian. Those skills were primarily for making daily life more comfortable and then once that was attained there was time for adding artistic embellishments to practical tools for its ascetic pleasure only.

However, when one does not have a manual on, say, 'how to make a Bronze Age funerary urn' how does one approach reconstructing it? When I started attempting to replicate the daily life of prehistory I set myself some basic rules with which to approach any reconstruction be it ceramics or house construction.

Do not acquire any training in the discipline you are replicating such as ceramics, textiles, metallurgy etc. If you have even the most basic training in say ceramic production you cannot help but be influenced by it. The only way to truly research a subject is to look at the artefact and work it out oneself. Then through trial and error one is likely to follow a similar route of discovery as our ancestors did.

Make sure there is a good attention to detail using only the materials available during the various prehistoric periods. It is no use making a log boat out of a tree that did not exist in the region at the time and expect the wood to behave in the same way as it did to the prehistoric boat maker. Making sure that the right materials were available and, also, the fabric of the tools used are essential to start with. Too many people that make reconstructions of all kinds think that as long as a material looks similar to the artefact it does not really matter if it comes from another part of the planet. A classic example of this bad practice was the Ice Man's cloak exhibit I saw at the Newgrange Museum in Ireland a number of years ago. Having made the replica of the 'Ice Man's' cloak for the museum where he is exhibited in Italy, I have a good knowledge of how it was made. However, it was not the technique that shocked me so much, it was the fact that it was made out of Raffia from the tropics and not the indigenous grasses used in the European Stone Age. This use of the wrong materials in prehistoric reconstructions is done all over Europe and I feel is the major reason that Experimental Archaeology has had a bad name with academics, justifiably. Another typical example of this practice I found whilst on an EAA (European Association of Archaeologists) conference excursion to Volos in Greece. Next to the museum we were visiting was a replica of a Neolithic hut. I walked over to take a closer look and found that where the daub had fallen from part of the wall there was steel construction mesh under the daub and plastic sheeting under the reed thatch. The modern hut makers obviously thought that such basic materials as wattle and daub and a reed thatched roof were not durable enough to withstand the weather. Yet such materials were used in prehistory and still are used all over the world today in house constructions quite satisfactorily.

Case study 1: Bronze Age Trevisker Ware ceramic production

I have, during my researches, made replicas of many types of prehistoric pottery and I always delighted in replicating the artistic embellishments that our forebears added to simple functional vessels. The discovery of ceramic I have long believed could have been achieved by accident whilst trying to make grass containers more durable and waterproof. When researching the possible uses of grass in prehistory I would make simple plaited and twinned grass bowls in various shapes and sizes. After a while, in an effort to make them more durable, I smoothed the inside of the bowls with clay, let it dry and then poured hot beeswax over them. This made quite an effective vessel in which to carry small amounts of water. A similar process was undertaken in the Fertile Crescent in the Pre-Pottery Neolithic where grass containers were lined with bitumen from the Dead Sea for a similar purpose. I once made a copy of such a bowl for a commission and, whilst very effective in holding water, it made the water smell of tar unless it was dried for a considerable amount of time before use.

Over time those clay lined grass bowls I made started to deteriorate and the simple solution was to throw them into the fire. If the fire was hot enough and the clay of ceramic quality, next day I would find in the ashes the internal clay shell of the basket turned to ceramic. On the outside it was beautifully decorated with the impressions of the plaited bowl it had once lined. A few years later I saw examples of ceramics that had been decorated on the outside to replicate a grass bowl too. This could have been an acknowledgement to that possible pre-pottery discovery. An example of this practice can be seen on a Bronze Age food vessel from Corbridge, Northumberland (Gibson 2002, 21)

In the Bronze Age in the south-west of England the dominant ceramic style was Trevisker Ware which comprised variously sized, elaborately decorated urns and vessels. This first case study hopefully throws a new light on a previously long-established assumption that Bronze Age Trevisker Ware pottery was made of gabbroic clay collected from the Lizard Peninsula. The Lizard Peninsula is on the south coast of Cornwall some 60 km overland from the Trevisker settlement, which was on the north coast. The landscape at that time was devoid of any road system having a densely wooded undulating topography.

Colin Renfrew suggested that the ceramics were made at the clay source. He says:

> whether they were produced by specialized potters living on the clay source or by potters coming seasonally from different settlements in the Cornish peninsula is not known. The high ratios of gabbroic pottery in settlement assemblages many miles away in north Cornwall indicates that substantial qualities were moved long distances (Renfrew 1977, 71–90)

In order to, hopefully, find these gabbroic ceramic manufacturing sites, the Lizard Project Landscape Survey was undertaken between 1978 and 1983 by members of the Cornwall Archaeological Society. Contrary to expectations, instead of finding large quantities of gabbroic pottery, ceramic manufacturing centres, or the remains of wasters from kilns, no production sites were found from this extensive survey and very few gabbroic pottery shards either (Smith 1987, 13–68).

It was suggested by Blackmore that 'the transport of raw clay over many miles is documented ethnographically and it is not impossible that inhabitants of settlements such as Trevisker and Gwithian might travel the 30 to 50 km by land, or even sail around Land's End, to collect raw clay' (Blackmore *et al.* 1979, 93–111).

These assumptions were feasible only if the local clays in the Trevisker area were unsuitable for the manufacture of ceramics. As a result of the survey undertaken showing there was no evidence of ceramic manufacture at the clay source on the Lizard, I felt it was a mystery that needed solving. Their arguments that wet clay was moved across great distances did not convince me because of the practicalities of moving wet clay. That is why most ceramic production is carried out at the clay source. I felt the only practical method of transportation, if it had to be moved, must have been in boats around the coast, as there was no traversable river from the Lizard peninsula to the Trevisker settlement. Primarily I needed to find out if there was a clay source at Trevisker to invalidate Blackmore's theory.

So I undertook an extensive survey in the Trevisker area, asking local farmers if there were any clay sources nearby. The farmers I spoke to were amazed that I should ask such a question, as clay in their area had severely hampered their farming practices. They informed me that the entire landscape was so clay rich that for generations their families had added tons of sand and lime to the soil to break it up so they could try to grow crops, but to no avail. Therefore, they had to restrict themselves to stock farming. The clay they showed me from the steam bank at Trevisker farm had such a high plasticity it could be taken directly from the ground and fashioned into a coil without any processing.

The clay was free of inclusions too, so by adding any kind of temper it could be made into substantial vessels. So why I wondered were the ceramic specialists convinced Trevisker Ware was made of gabbroic clay?

Michael Parker Pearson in his paper for Cornish Archaeology states 'the large gabbroic rock inclusions in many of the vessels made such identifications relatively simple' (Parker Pearson 1990, 5–32). So it was the large gabbroic rock inclusions from the source of the Lizard peninsula that had made them think it was made of that particular clay? Immediately I knew how this misconception could have been made. My previous experience of gabbroic stones from the Lizard peninsula was that they were used in prehistory for the purpose of heating water and cooking with. There is a wide misconception about the prehistoric period that they had none of what we would think of as the basic comforts of life, such as constant hot water. This was not the case, as igneous beach stones were always at the base of the central dwelling fires and, as a result, a constant supply of hot water was available whenever needed. All that was required was two forked sticks to get the stones out of the fire and drop them into a container of water. My own experiments showed that a container of water will heat in this way twice as fast as with a modern electric kettle. Once the water is hot enough the stones are then forked out and thrown back into the fire until needed again. No extra wood is required to heat the stones as the central dwelling fire would be constantly lit.

The use of firestones for cooking in hunting camps known as *Fulacht Fiadh* in Ireland is well documented and identified in the archaeology by crescent shaped piles of fire cracked stones, or burnt mounds, as they are known. A survey of the longevity of various stones used in this practice was conducted by Victor Buckley who says 'whereas many types of stones can be used for this purpose it was found that gabbroic stones were by far the best and can be used more than 25 times before they begin to crack' (Buckley 1990, 170–172)

This practice of heating water with hot stones was an important part of daily life in prehistory, so a source of these particular stones would have been a valuable trading commodity, even in Cornwall with its abundance of granite stones that could also be used for the same purpose. Granite stones can be heated about eight times before they crack quite dramatically whereas, even after 25 times, the broken gabbroic stones can still be used quite effectively. However, even those broken stones eventually become too small to be of any use, so they could then be crushed and added as a temper to clay. The abilities of this particular stone to withstand the thermal shock of heating and quenching would also have that same ability when added to the plastic clay of Trevisker for bonfire firing, especially when firing large vessels during damp weather.

As the Trevisker farmers today with modern farming techniques still cannot grow grain on their land I doubt very much if the Bronze Age farmers could do so, yet grain was excavated at the settlement. So I came to a conclusion about a possible scenario.

A boat could pick up a load of gabbroic stones from the Lizard peninsula for heating water and a small amount of gabbroic clay as an immovable ballast in the boat in which to set the stones for the journey to circumnavigate the north coast swell. If the stones were not set into wet clay in the boat and a wave hit it from the side, the round beach stones would roll with the wave and could capsize the boat. As the vessel followed the coast around Lands End it could pick up grain from settlements for the Trevisker ceramic manufacturers. Once there, the stones, clay ballast and grain were deposited (Fig. 4.1).

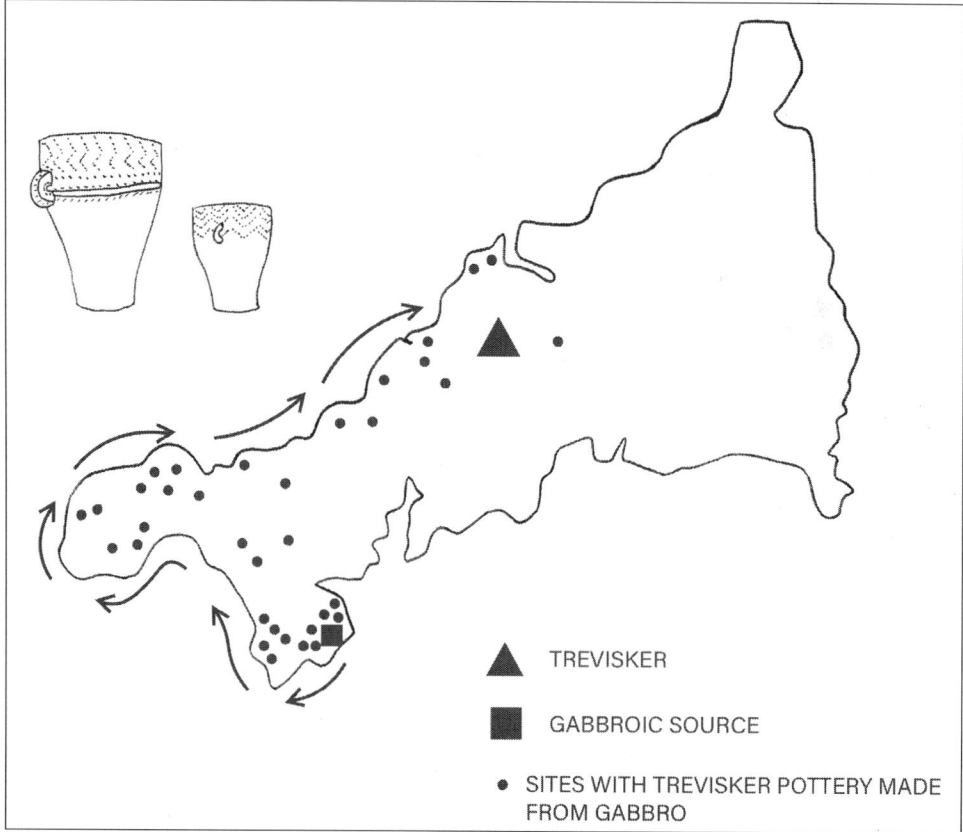

TREVISKER

GABBROIC SOURCE

● SITES WITH TREVISKER POTTERY MADE FROM GABBRO

Figure 4.1: Map of Cornwall showing the route from the Lizard and Gabbroic source and around the coast to the Trevisker settlement. Marked are settlements that had Trevisker Ware in the archaeology

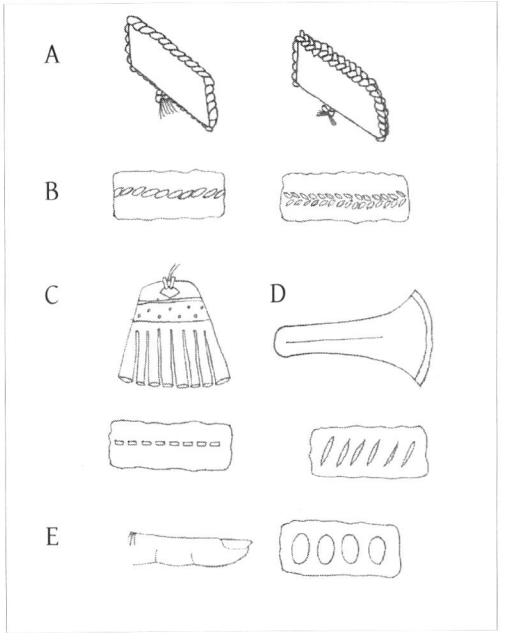

Figure 4.2: Tools used to make Trevisker Ware ceramic decorations: a) plait cord; b) twisted cord; c) comb; d) scored; e) finger print

On the return journey Trevisker clay could be put in the bottom of the boat and the Trevisker ceramics set into the wet clay to stop them rolling and being damaged on the return journey. The ceramics were then distributed to the settlements that had provided the grain for the north Cornwall ceramic producers.

I needed to attempt to make some Trevisker ware ceramics to see if they looked like the ones excavated with the large fire cracked gabbroic inclusions in them. I decided to make three copies of large Trevisker Ware funerary urns to test this hypothesis. I planned to read a paper on my findings at the EAA Lisbon conference in September 2000.

So I decided to make the urns with the coil method and then had to replicate the distinctive Trevisker style decorations on the urns. At a Bronze Age site in Cornwall called Trethellan, situated on the coast south of Trevisker, a wide variety of Trevisker type pot decoration can be found on the ceramics (Fig 4.2, a–e; Nowakowski 1991, 109–120).

Finding fire cracked gabbroic stones was not a problem as I had been using those stones for cooking at my research centre for years. I crushed some into very fine particles leaving some relatively large pieces to replicate the shards in the Trevisker pots in the archaeology. I then mixed the plastic Trevisker clay with approximately 40% of the mixed crushed gabbroic temper. This made it very gritty clay to coil but the plasticity of the Trevisker clay helped it hold together well enough for making such a large vessel. I made the pots in the spring thinking they could have all the summer to dry out naturally before firing. However, it was a very wet summer in Cornwall that year and I kept putting off the firing until they had dried out completely. The weather did not improve, so I just had to take a chance that the urns were not too damp to be fired in an open bonfire.

Firing method

1. I had collected a large pile of dried blackthorn (*Prunus spinosa*) , a scrubby hardwood prevalent in Cornwall, and laid the urns on their sides on a grassy field next to a slope for some updraft (Fig. 4.3, a).

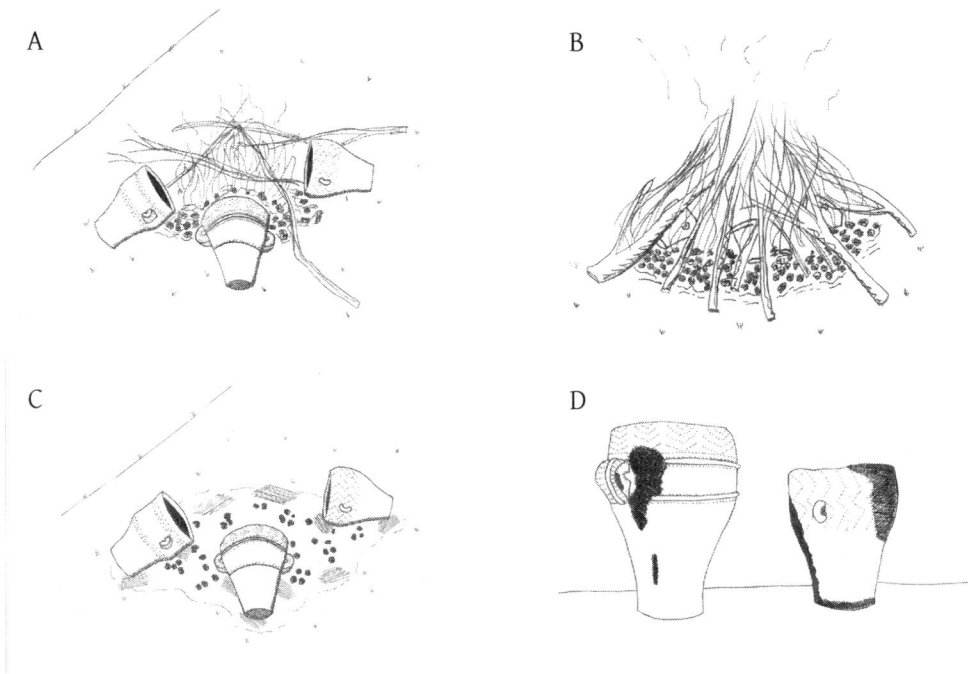

Figure 4.3: a) woodpile and three urns set on grass for firing; b) fire enveloping urns; c) urns fired; d) fired urns with burnt grass markings

2. I lit a small fire at the front of the pots, then I enveloped the pots in the fire and piled more wood on top (Fig. 4.3, b).
3. I assumed due to the size of the pots I would have to fire them for at least three or four hours, but after one hour a stick fell onto one of the pots and the pot rang like it had already turned into ceramic, so I let the fire go out (Fig. 4.3, c).
4. What I found in the ashes were three perfectly fired urns with the distinctive black marks on their sides so often seen on prehistoric ceramics in museum cases. This mark was a result of the burnt grass they were resting on during the firing process (Fig. 4.3, d).

This exercise was an example of my self-imposed restrictions of not having any training in any of the artisan craft skills I was trying to replicate.

In 2010 the Dartmoor National Park Authority and the Royal Albert Museum in Exeter commissioned a professional potter, Joss Hibbs from the Powder Mills Pottery on Dartmoor, to make a replica of Trevisker style Urn for a forthcoming museum exhibit. Ms Hibbs, being a professional potter, could not help being influenced by her years of experience firing pots in a wood fuelled kiln. After making the pot she had prepared two tons of split wood for the pot firing and set about making a turf kiln. She built the kiln as a replica of the sort she was used to using, with a fire box, etc.

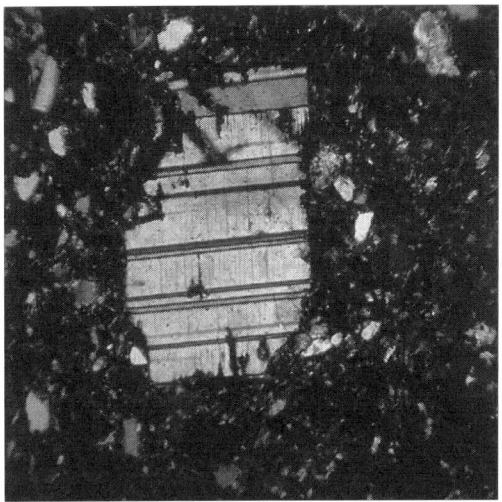

Figure 4.4: Thin section of Trevisker Ware shard with thermally cracked gabbroic inclusion in a background matrix not from Lizard peninsula (photo: James Strongman, Petrolab)

This was so that the heat would build up slowly in the kiln around the pot. In her blog she states that she must get the pot to 600°C before she could let the flames touch the fabric of the pot itself, or it would crack. She fired the pot for 10 hours and then sealed the turf roof of the kiln and left it for 36 hours to cool, before breaking it apart and getting the urn out. The pot she made was fired successfully but it was not bonfire fired, it was kiln fired in much the same way as pots have been fired since the Roman period.

My urns were laid on a grassy slope in the open, the fire was lit in front of them and they were immediately enveloped in the fire, once the fire took hold. It took 1 hour using sticks from local scrub trees to fire them and immediately, once cool enough to handle, the urns were picked up ready for use. The amount of split wood, turf kiln making and the 46 hours firing and cooling of the pots is a prime example of how knowledge of any modern artisan skill impedes, rather than helps, the artist when trying to replicate prehistoric skills. It is the large amount of crushed gabbroic rock in the clay that helps the ceramic to withstand the thermal shock of heating up a damp vessel quickly and equally quickly cooling it down.

So, my urns successfully fired, I presented my paper at the Lisbon EAA conference and it went down very well with the assembly. On returning home I wrote an article for *British Archaeology* magazine on my findings (Wood 1999, 8–11). In the following issue however, there was a letter from someone who had been involved in the analysis of Trevisker Ware ceramics in Cornwall. She said, more-or-less, that whilst the crushed firestones as a temper was a good theory, there was no evidence for it in the archaeology. She said she had been involved in a project that had conducted hundreds of thin sections of Trevisker ware ceramics and had not found one with thermally cracked gabbroic rock in them (Harrod 2000, 24).

I was disappointed by this and left my theory for a year or so, but I still thought my assumption was valid. So I decided to get some samples of Trevisker Ware from the Royal Cornwall Museum store and have some of my own thin sections done to see if I could find any thermally cracked gabbroic inclusions in them. When the report of the samples came back from the local geology company that analysed the thin sections it showed I had found my evidence after all. Whilst some of the samples were made up of an add mix of gabbroic clay and a north Cornwall clay, there was also a sample of a thermally cracked gabbroic rock inclusion in a clay matrix that was definitely not from the Lizard Peninsula (Fig. 4.4).

So this validates my theory that, if an established concept about a prehistoric practice from any discipline seems illogical, and impractical, then it most likely is.

Case study 2: making prehistoric mountain attire

In 1998 I was commissioned to make a replica of a grass cloak, shoes and dagger scabbard for the Ötzi exhibit at the South Tyrol Museum of Archaeology in northern Italy. Having spent some time at the museum studying the Ötzi artefacts prior to making copies of them, I always thought the display of his back pack was curious. I could not see how he could have worn a back pack over his grass cloak at the same time as his bow and quiver (Fig. 4.5).

However, I was not commissioned to make the back pack, so I just concentrated on making the cloak and shoes, etc. Whilst making the shoes I did notice an anomaly, which was a detachable strip of leather under the sole of the shoes. I assumed at the time it must have been for a ski or snowshoe of some kind but did not think that would be unusual for an alpine dweller. Konrad Spindler, leader of the scientific excavation of Ötzi, suggested in his book 'two leather straps knotted under the sole leather were designed to ensure a better grip' (Spindler 1994, 179).

I thought that might have been a valid assumption if the leather strap was a thick and an angular piece of leather. It was, however, flat and thin, so it would not be an aid when walking on icy ground at all. It was not until years later when I was working with a TV company as a consultant for a BBC documentary on how Ötzi might have died that I came to my conclusion about the back pack. I was at the meeting in London with one of the directors and happened to mention that I thought something must have been attached to his shoes like skis or snowshoes. One of his researchers decided to look for snowshoes on the internet while I was there and found hundreds of old American Indian snowshoe frames for sale to be used as wall decorations.

Then she found this picture and as soon as I saw it, I realised that Ötzi could have had part of one of his snowshoe frames with him when he was found

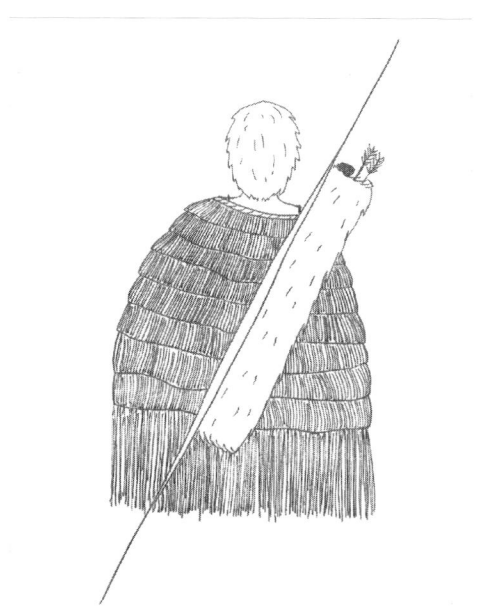

Figure 4.5: Ötzi museum exhibit with bow and quiver over his grass cloak

(Fig. 4.6, a), misinterpreted as his back pack (Fig. 4.6, b). The TV company liked my conclusion and commissioned me to make a pair of Ötzi snowshoes for the film.

First I made the basic shoes as I had done before, slits were cut around the bear skin sole to accommodate the strips of red deer skin that would hold the lime bark twine to the soles. I realised when I first attempted to make the shoes that the bark twine was not a pre-made 2-ply twine but twisted whilst making the shoes. At first I could not see why they did not just make a ball of 2-ply bark twine and then just attached it to the leather strips through the slits in the soles.

So I threaded a strip of un-twisted lime bark through the deer skin strip and twisted it together up to the height of the ankle (Fig. 4.7, a).Then I twisted them together for 1 cm, as they were on the artefact, ready to take the bark strip down again to the sole. Suddenly it became obvious why the twine was not pre-made before making the shoe (Fig. 4.7, b). If you held back the same half of twine at the ankle and let the other half go down to the sole and up to meet it again, you would create the equivalent of a draw string at the ankle. This would mean that when the net was finished the other half of the twine could be pulled to fit snugly onto the foot (Fig. 4.7, c).

The next step was to cut and twist the cross members of the bark to make the net to accommodate the grass insulation (Fig. 4.7, d). I have over the years worn copies of these shoes in the summer without the grass insulation or deer skin panels and they made very comfortable sandals.

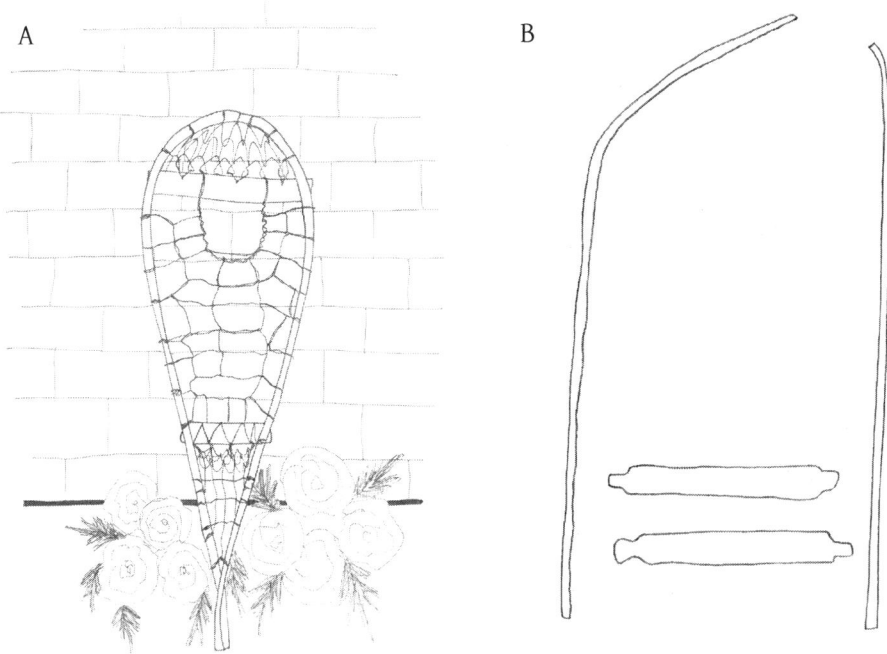

A   B

Figure 4.6: a) Internet advert for old American Indian snowshoe; b) back pack frame and larch slats

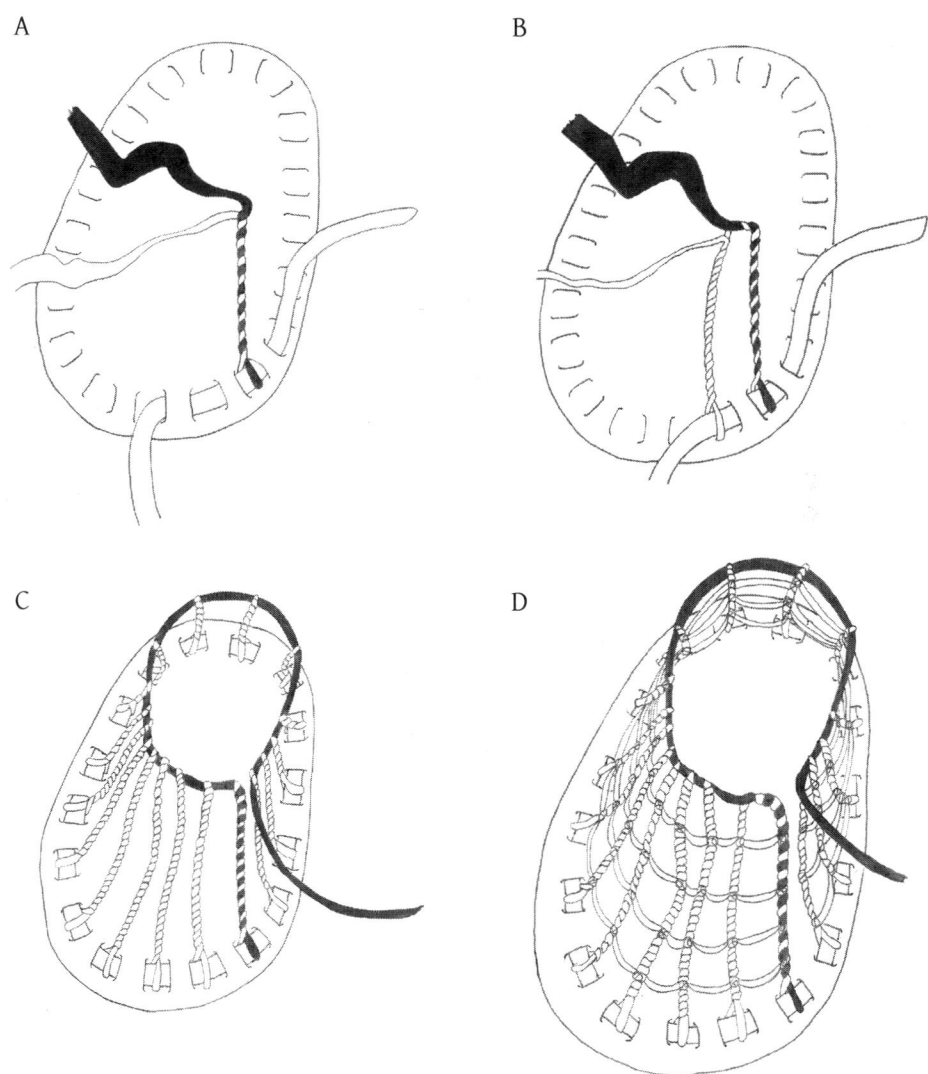

Figure 4.7: Process making bark string net to shoes: a) first strand; b) two strands showing the drawstring; c) completed drawstring around ankle area; d) cross strings to hold in the grass insulation

The deerskin panel was then attached to the front of the shoe by adding another thin strip of red deer skin looped through the slits that accommodated the bark twine for the net part of the shoe.

So I decided to look into the snowshoe concept in more detail and at the same time resolve another burning question I also had about the shoes. This was why in the well-known Mainz Museum sketch of the shoes had no leather backs on them over the bark netting. If one walked in the snow with shoes with only a grass stuffed string back to them your feet would quickly get very wet and cold. Added to that, the grass stuffing of the shoes, presumably for insulation from the cold, would have become like a sponge holding the melted snow around the feet. Ötzi was so well protected from the elements, with his thick fur coat topped with his grass cloak as insulation, it made no sense that his shoes were so inadequate for the alpine winter climate. I looked again at the archaeological plan of the shoe remains that were found next to his body and it indicated that there were more slits and soft leather inserts not accommodated in the Mainz Museum interpretation, showing how the net was completely covered in leather (Fig. 4.8).

In order to attempt to make the snowshoes I did a survey of over a 100 American Indian snowshoes for sale on the internet. There seemed to be generally three sizes of snowshoe frame 42 × 14 inches, 36 × 11 inches and 33 × 10 inches (106.68 × 35.56 cm; 91.44 × 27.94 cm; and 83.82 × 25.4 cm); the first measurement was the length of the wood of the snowshoe frame and the second the width of the frame itself. The three sizes were presumably made specifically for the height of the wearer. The snow-shoes required for the height of Ötzi who was 5 ft 3inches (1.6 m) tall had a frame of approximately 36 inches (91.44 cm) long. This was the same size as the 'back pack' frame giving validity to my conclusion that it was not a back pack frame at all. The two strips of larch wood thought to be the base of the back pack were just the right size to tie across the snowshoe to support the foot.

In the archaeology, Konrad Spindler also observed 'numerous cords found by and on the ledge, and especially in the crevice; mostly they are torn into fairly short pieces. "Any amount of string" was Messner's comment on the find. Proof is provided by the impressions of cords on the hazel rod. We interpret the constructed form as a frame of a back pannier' (Spindler 1994, 92–93). This clump of knotted twine excavated from the ledge near the body of Ötzi with impressions of the cord on the hazel rod was more likely to be the knotted twine mesh needed to make the snowshoe.

Putting these observations together and the thickness of the string, which was 6–7 mm thick once plied, I proceeded to make some string to construct my replica.

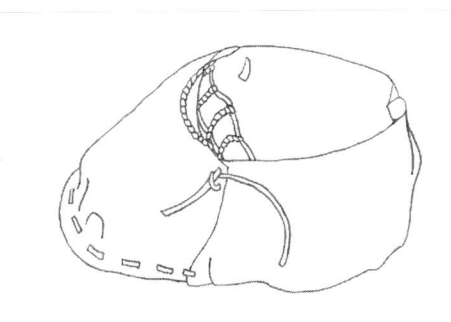

Figure 4.8: Leather back panels on shoe to attach to the snowshoe

I made the string out of *Tilia cordata* or the bark of the small leafed lime tree. Ötzi had this bark string as the net for his shoes, the cross members of his grass cloak and his dagger scabbard. It took approximately six hours to make this string for one snowshoe.

I then made the snowshoe frames (Fig. 4.9, a & b).

Using the information from the plan of the strips of leather I found it rela-tively easy to find out how they were attached to the larch slats that were the

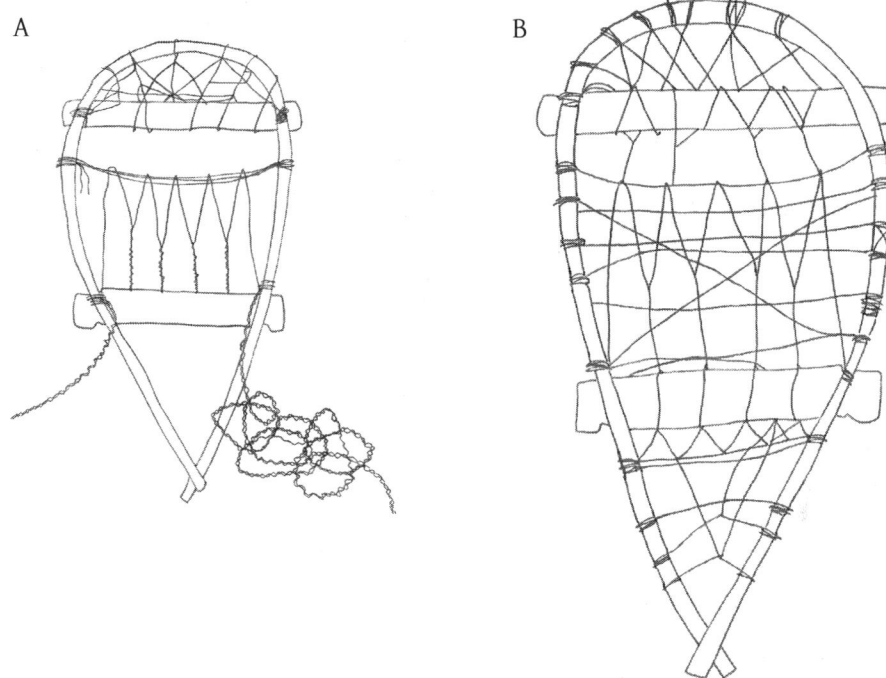

Figure 4.9: a) Larch laths attached the frame and bark string net started; b) completed snowshoe

cross members of the frame. Those straps also had slits in them to accommodate the bark twine to pull it tightly to the shoe and secure it.

Amazingly the strap that made me think originally it was a snowshoe was not actually attached to the snowshoe frame at all. It was only when I had almost attached the shoe to the frame that I realised I had not used the strap from under the sole. So I undid one side of it and found it went over the front panel of the shoe then it could be attached to the other side of the sole (Fig. 4.10, a). This fitted so perfectly it took me a while to realise that it was to accommodate the remaining shoe ankle twine to firmly keep the front of the shoe attached securely to the snowshoe frame (Fig. 4.10, b). Even though this strip was not used underneath the sole it was the most important piece of leather because without it the front of the shoe would lift from the snowshoe frame whilst walking.

The grass cloak

The practicality of wearing a replica garment has always to be addressed because there is no validity in making something for a manikin for a museum exhibit if it is not practical to wear. While I was making the replica grass cloak for the museum I suggested to the museum director that I felt there had to be side slits in the cloak. The museum preferred to not have slits as it would not fit well on the manikin that

A

B

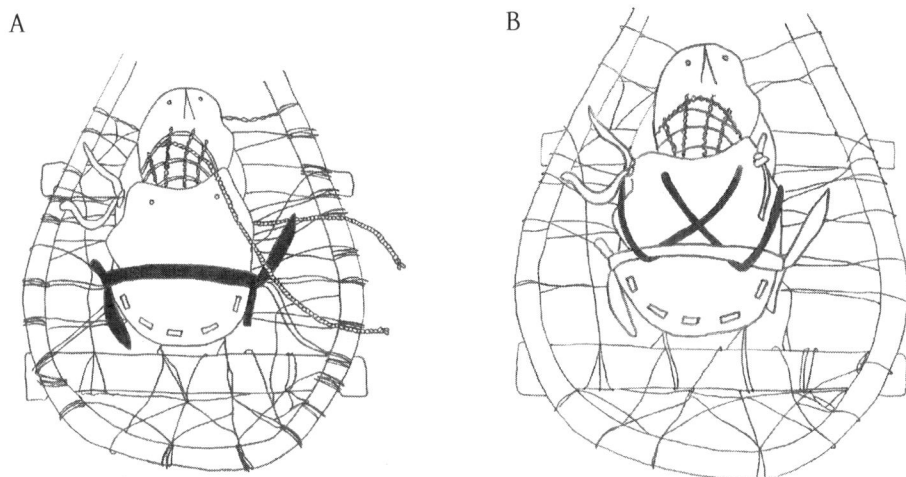

Figure 4.10: a) Front panel of shoe with underside strap attached; b) how the string from the ankle of the shoe loops through the strap and onto the frame to keep the foot from lifting when walking

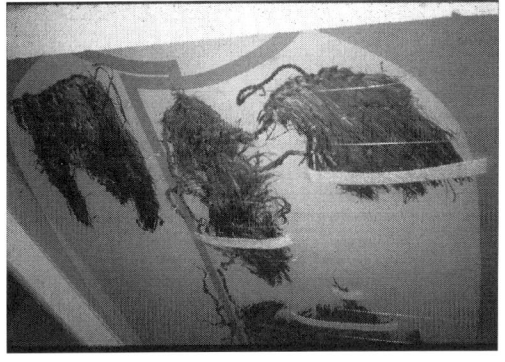

Figure 4.11: Ötzi's grass cloak taken before the museum opened in Bolzano (photo: Jacqui Wood)

was specially made for the exhibit. On the artefact there was a series of loops down the front of the cloak and a long cord at the neck indicating it was to lace up the front of the garment through the loops, to keep the chest warm and dry (Fig. 4.11).

If this was laced up without side slits the cloak would, in fact, make a straight jacket and be unwearable. If the slits were inserted and a piece of cord tied around it at the waist, it made the cloak a manageable and practical garment to wear (Fig. 4.12, a). Another issue I had whilst making the replica was the lack of a top to the grass cloak itself. When inspecting the original artefact prior to making the replica I noticed the cloak must have been worn off the shoulder and was held in place by clearly defined shoulder straps (Fig. 4.11).

The reason that the cloak had to be made in this way was because, in order to plait in enough grasses to make the cloak thick enough to repel the elements, the plait itself had to be a certain length. This was an anomaly because whilst Ötzi was very well protected from the elements from his shoes to his bear skin cap, his neck and shoulder area was completely exposed to rain and snow. This did not make any sense, so I devised a small grass cape that could have been worn over the top of the cloak in much the same way as a riding- or raincoat has an extra cape attached at

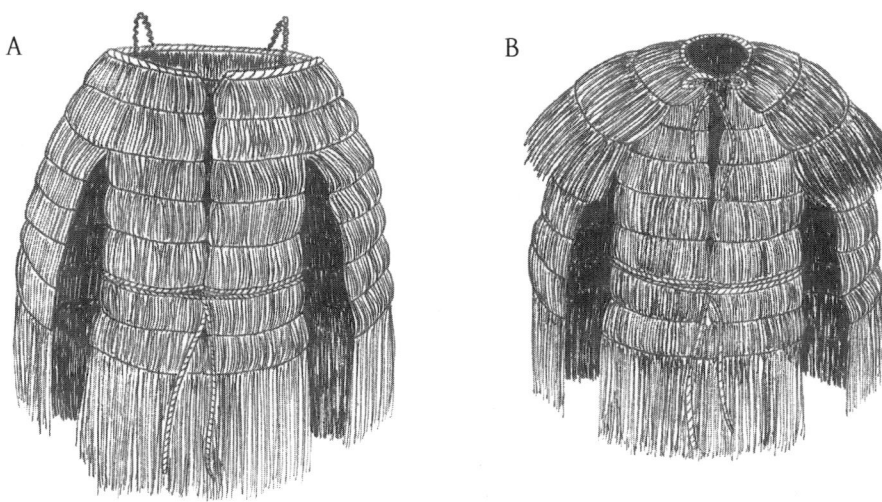

Figure 4.12: a) Grass cloak laced up with bark belt holding it close to the chest; b) grass cloak with additional cape to protect shoulder and neck area from the elements

the shoulders (Fig. 4.12, b). There is no way of knowing if he did wear such a cape (unless of course it is found) but, due to the thickness of the long off the shoulder plait, his body could not have been covered by one grass garment. Therefore, it is not inconceivable that this addition was worn to complete his weather proof outfit (Wood 2004, 119).

In conclusion

The anomaly of the strap under the shoe suggested that something had, in fact, been attached to the underside of the shoe. Conversely that strap was, I believe, to hold down the top of the shoe and not to attach itself to the snowshoe frame. It was, however, the vital part that made it possible to wear the snowshoe at all. The similarity of the shape of the frame of the back pack and the shape of a typical American Indian snowshoe cannot be denied. Added to this, the impossibility of wearing a back pack over a cloak with his bow and arrows (Fig. 4.5) gives validity, I feel, to my theory: which is, that Ötzi did have snowshoes with him when he went into the mountains and not a back pack. Conversely, the main objection to my snowshoe theory is that there were two slots found at either side of the base of the hazel wood frame. The larch slats used in the snowshoes fit into those slots giving birth to the back pack concept. However, this does not get around the reality that you cannot wear a back pack over a cloak in the first place. I suggest that the slots at the base of the hazel wood frame were for utilising it into a back pack for summer use only, when the grass cloak would not be worn. Very much in the same way as I mentioned earlier in this chapter when I utilised the shoes by removing the leather panels and the grass insulation to make them into very functional summer sandals.

Also, the grass cloak was clearly made to be laced up at the chest, which would have made it impossible to wear without slits in the cape. In addition, it is inconceivable that Ötzi had no protection from the elements at his neck and shoulders when he was so well protected from the cold and snow with the rest of his attire.

The only way one can come to such conclusions is by following the practical steps the artisans must have taken in prehistory when making his mountain equipment and clothing.

Case study 3: the prehistoric architect

The first house I was involved in the re-construction of was copied from a Gwithian house plan from the Bronze Age, excavated in Cornwall in the 1950s by Charles Thomas. Basically all we had to go on was the site plan of a 7.6 m diameter roundhouse with nine ring-beam posts making an inner circle and an outer circle of posts indicating the outer wall was made of wattle and daub.

Various centres at the time (in 1992) were building roundhouses after going on a 'prehistoric building course' at Butser Ancient Farm in Hampshire. Dr Peter Reynolds, who ran the centre, was a pioneer in this discipline, so it seemed the obvious place to go and get some instruction. However, I really did not want to see a reconstruction of a prehistoric house until I had tried to make one myself. So I looked at the archaeology for inspiration, not expecting to find much until I came across a ceramic house urn dated to the 7th century BCE from Königsaue in Germany (Fig. 4.13) which indicated the house had a very steep roof pitch (Audouze & Buchsenschutz 1992, 83). There were also a few drawings of roofs at the Val Camonica rock art site in northern Italy, showing a similar roof pitch to the Königsaue ceramic.

So, with a site plan, no funding and the land to put the house on we needed to find a way to get the materials to make it. Those for the wattle and daub outer wall were easy to find on my own land. There was enough silty clay in the stream banks to mix with soil and straw for the daub and more than enough hazel and willow for the wattling. Finding long straight poles for the roof frame, however, we thought would not be so easy to acquire. I did not want to use farmed pine poles (*Pinus*) which were easy to acquire but were not part of the Bronze Age landscape. I wanted

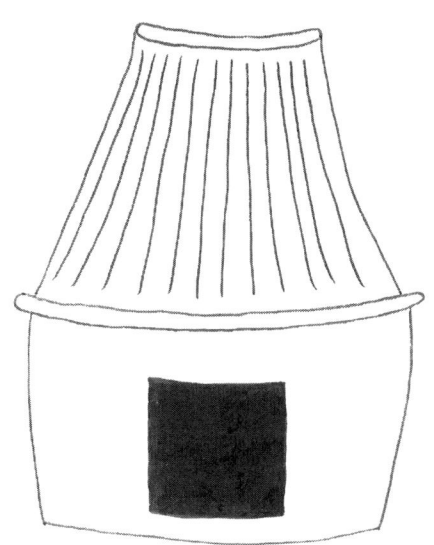

Figure 4.13: Königsaue House Urn c. 7th century BCE

to try and get the wood our forebears would have used, such as ash or alder which would have been readily available in Cornwall at the time.

It was suggested by a friend that if we followed the electricity cable runs that crossed local woodland there might be a crop of long coppiced poles readily available. If a cable run crosses a low-lying valley the electricity company have to pay woodmen to cut down the trees every few years so they do not grow too tall and touch the cables above. Once we started looking for such cable runs it was relatively easy to get permission from the land owners to cut them and take them away. The electricity company was happy with the arrangement too, as it saved them having to pay people to cut them down. The larger ring-beam posts however, were found on my own farm land, so all that was needed was water reeds for the thatch.

All the centres around the country bought reeds to make their reconstructions but, with funding being non-existent, this was not an option for us. So we decided foraging for our reeds in the country would give us insights into the various problems our ancestors might have encountered when they searched for roofing materials. We found three very distinct varieties of water reeds (*Phragmites australis*), one *c.* 2 m long and thick like bamboo, one very fine and no more than 1 m long and one as fine as the 1 m long reeds and yet a good 2 m long, which were perfect for the roundhouse. The short variety was eminently suitable for thatching cottages, being a similar length to wheat straw but twice as durable. However, those same short reeds needed three times the number of batons to tie them onto a roundhouse, making the whole process of thatching much more time consuming. The tall thick ones had their uses too in making warm springy bases for beds or seats. They were, however, no good for thatching as they broke very easily and were also harder to tie tightly onto the batons to secure them to the roof.

The county of Cornwall had quite a number of reed beds and the largest was owned by the RSPB (Royal Society for the Protection of Birds) as a bird sanctuary. The need for cutting reeds for thatching cottages in Cornwall was superseded in the 17th century by the quarrying of slate at Delabole in north Cornwall. Therefore, all the Cornish reed beds had been long abandoned to the wildlife. Coincidentally, when we first started looking for reeds in 1992 the RSPB began a new policy to cut a quarter of their reed beds every year as the fresh growth benefited some species of birds. So, when we asked if we could cut some and take them away for educational purposes they were delighted. After gathering a group of volunteers who were willing to walk on a floating mass of reed roots over the edge of a lake in mid-winter, we had our roofing material. The only time we could cut the reeds, however, was between December and mid-February. The reason for this was that it took until December for the lower leaves of the reeds to decompose, leaving just the thatching material. We tried to thatch another house years later with reeds cut in September that still had the soft side leaves attached. We found that, whilst it looked like it made an effective roofing material, the rain in the winter tended to soak into the roof like a sponge. The soft side leaves did not shed the water and within a year the thatch had to be taken off as it had gone mouldy and leaked. The reeds could also not be cut after mid-February because there was a danger we might disturb early nesting birds.

We began building our first roundhouse with nothing but the raw materials and a rough idea of what the house should look like. When constructing a wattle and daub roundhouse the first question is always: how high were the outer walls? The archaeology gave no indication as to the height, just the spacing of the post-holes. We could have built the walls no more than 60 cm high, making the roundhouse more or less teepee shape with a little extra height at the sides. The roundhouses at Butser Ancient Farm had 1.6–2 m high walls but there was a reason for that. They employed a cottage thatcher to help them thatch their roofs, so they replicated the 45% angle of a cottage roof and made the thatch at least 30 cm thick. This must have had something to do with their decision to make the wattle walls so high as the roof pitch was not steep and if the walls were low there would be very little standing room inside the dwellings.

As our roof pitch was much steeper, 55% replicating the Königsaue ceramic, we decided to make our walls 1.20 m high as the roof was steep enough not to restrict walking along the inside edge of the dwelling. Deconstructing the process used to build a prehistoric dwelling or any artefact I found over the years is a very intuitive process. One has to put oneself in the mindset of 'what would I do in that situation' just as you might have to do if you were washed up on the proverbial desert island and needed to make a shelter. After 27 years of experimental archaeological research I have found this is the best way to approach any project.

- Do not attempt to replicate anything, be it textile or house, with a rigid plan of how to do it that you are going to stick to.
- One must have the plan or artefact and an awareness of the right materials for making them and then just start the research with only a vague idea of what one is going to do.
- Then, as I have always found once into the process, it becomes obvious there is only one way to do the reconstruction and it is that fluidity of mind that puts you in the mindset of our prehistoric ancestors and it is that mindset where the groundbreaking research is done.

The hut we were hoping to replicate was round, so how would we make a perfect circle without any modern equipment? After much thought we found that a piece of string, two sticks and a stone with a hole in it was all the equipment that we needed to be a prehistoric roundhouse builder. The construction sequence was as follows:

1. Drive one of the sticks into the ground, tie the string to it and tie the other end of the string to the other stick. Then draw a circle in the ground keeping the string tight (Fig. 4.14, a).
2. Drive in the posts for the wattle walls on your drawn circle. After putting in the posts, weave long willow withies or hazel between the posts (we found hazel the

best but willow a close second to do this job). As we decided to build our walls 1.2 m high this needed to be completed first allowing for the typical south easterly opening for a doorway in the circle (Fig. 4.14, b).

3. The nine ring-beam posts, which were 18 cm in diameter, must then be set into the interior space. The Gwithian house had its ring-beam set 1 m in from the outer walls. We dug holes 46 cm deep for the poles to be set into and then packed them well with stones. The cross-beams were then added and lashed at the top of the ring-beam posts. On this first house we cut the nine ring-beam posts with thick natural forks at the top of them. This took considerably longer to find in the forest but we thought it would make attaching the ring-beam to them a relatively easy process, which it was (Fig. 4.14, c).

4. Once the ring-beam was completed we needed to put the rafters on the roof. Having cut a number of ash poles from the woods we tied the three tallest and straightest at the top with our piece of string, having untied it from the central stick in the middle of the floor. On the other end of the string we tied the stone with the hole in it. Three people were then selected to hold the other end of the rafters and walk slowly around the outside of the wattle wall. When it seemed they were evenly spaced apart they held their rafter next to the nearest post in the wattle fence. The next process was to move the rafters up and down until the string with the stone attached was directly over the original stick that was driven into the middle of the floor. Once the stone was directly over the stick the rafter holders tied their rafter to the post they were next to. Then more rafters were added and tied to the nearest post until the roof frame was complete. The reason for having a ring beam in the roundhouse was so that the green ash rafters would season *in situ*. At a roundhouse reconstruction in Wales at Castell Henllys, they made the mistake of interpreting what they thought was a house plan with one that must have been an animal pen of some sort. They thought this 'house' did not have a ring-beam and built it accordingly. After they constructed the house and attached the heavy thatch to the rafters they thought the roof was sound. However, during the first winter, the rafters all dipped in the middle as they were bent out of shape by the weight of the thatch because the rafters were green wood (Fig. 4.14, d).

5. The next process was to attach the batons to the rafters with 2–3 cm diameter hazel sticks. These batons were attached in rounds every 30 cm and became the ladder for the baton tiers to climb up on the frame to the top of the house as they worked. The frame of the roundhouse was then complete and ready for thatching with reeds which were tied in layers until they reached the top (Fig. 4.14, e).

We imagined we would never get enough reeds from the marshes to make a thick thatch in the first season. We thought if we could at least cover it with a thin layer of reeds *c.* 12 cm thick, we could add another layer the following season. This is where research done with no funding can prove to be more insightful than just ordering tons of reeds from commercial reed suppliers. We had inadvertently made

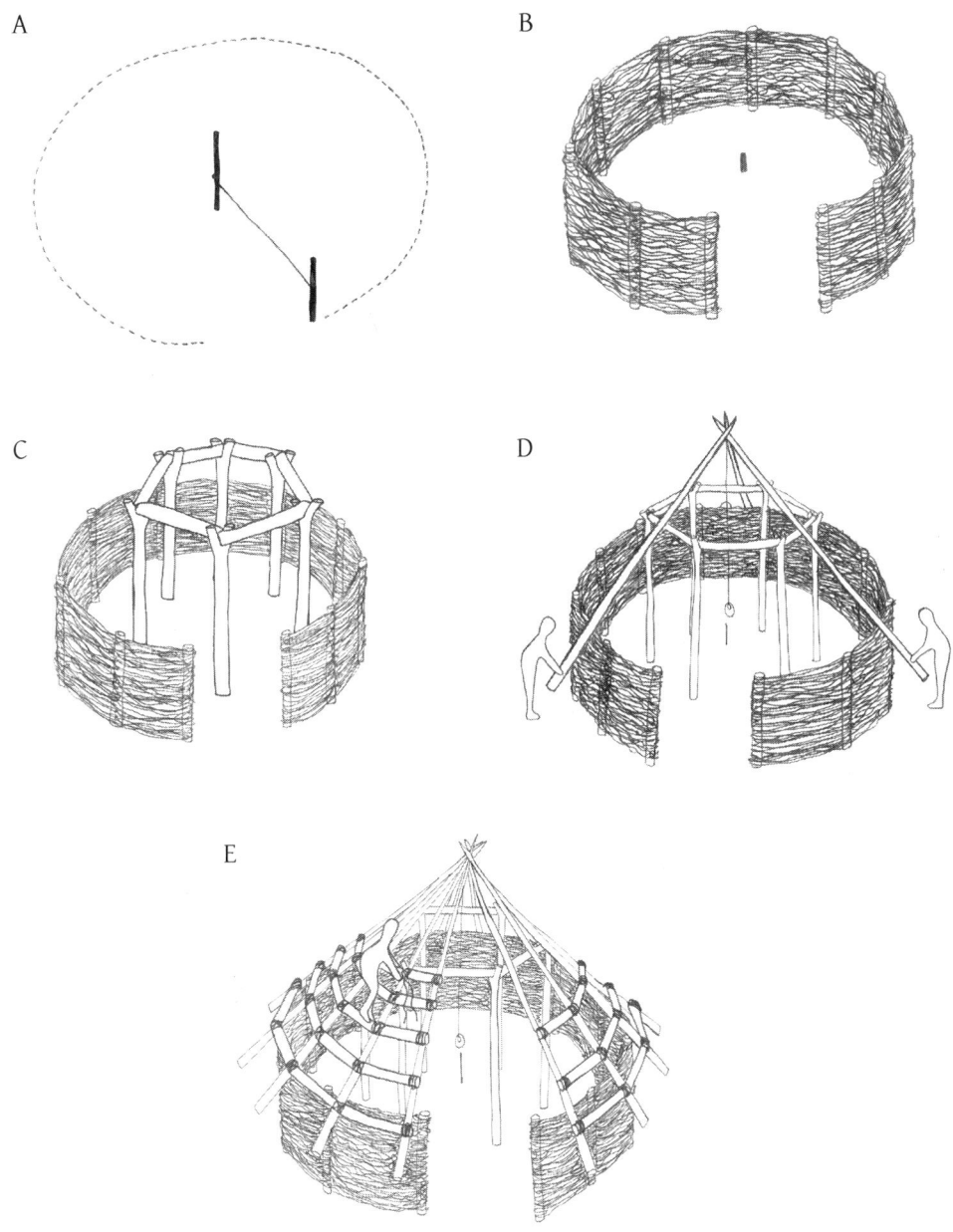

Figure 4.14: a) sticks and strings making circle to make line for roundhouse wall; b) wattle walls made; c) wattle walls and ring-beam in place; d) three people positioning first rafters over central peg stick; e) rafters on and first layers of batons tied

a major discovery by only adding that thin layer of reeds. We found that our steep, 55% pitched roof inspired by the Königsaue house urn, was designed for a thin layer of reeds. When it rained, because the pitch was so steep, the rain swiftly ran off it before seeping through. More importantly though, the smoke from the central fire went straight through it because it was not too thick. Having taught prehistoric cooking techniques in many roundhouse reconstructions that were thickly thatched, I found them unbearably smoky. Once the fire was lit the smoke filled up the roof space and then started to pour out of the doorway assaulting the senses of anyone entering. A lit central fire would have been a constant feature during prehistory thus making an environment of constant smoke, especially if damp wood was used, as it would have been during the winter. I found it unbearable to work in those constantly smoky conditions, to say nothing of its effect on one's health; so much so that I regularly had a fire made outside in order to complete my cooking courses. It is inconceivable that anyone would live in such a smoky environment for more than a day or two.

This is an example of making a reconstruction that looks good to get the fee paying public to such centres, as long as one does not light a fire in the house. Some of the very large Iron Age reconstructions of roundhouse are fine with such a thatch though because their roof spaces are so vast they can accommodate the volume of smoke. The average dwelling in the Bronze Age was approximately 7 m in diameter and just would not work with such a thick thatch. One might think that a hole could be cut into the top of the roof like a chimney in a cottage to solve this problem. This, however, was tried and it was found that it drew too many combustibles up into the thatch and a number of roundhouses burnt down as a consequence. Our houses were never smoky and when we eventually took down the original house, after having 15 years of fires in it, the thatch on the inside of the roof was as golden as the day we put it on.

After we built the roundhouse I wanted to make an area inside for weaving textiles and yet even on a bright sunny day it was still quite dark inside. Weaving would have been a winter activity when it was not possible to work outside. With the short winter days it would have been almost impossible to weave fabric in such darkness. I decided to cut windows into the wattle walls around the house and made reed and skin blinds to drop down in the evening or if it was windy. This seemed to me to be an obvious thing to do, yet I could not, at the time, find any evidence of this practice in the archaeology. Doing this sort of project always makes you think, what would I do in that situation? Cutting holes in the wall was a simple solution and made the dwelling much easier to work inside. There is a lot of ethnographic evidence around the world for windows in wattle walls to substantiate this theory.

What one has to remember is that people in the Bronze Age would have had the same logical mindset as we do today, wanting much the same basic needs whilst living in such a dwelling. So what a practical person might think would be a good idea, is very likely what would have been done then.

An interesting thing I found was that only when one replicates such buildings does one have real insights into the wider consequences of the building process. For instance, I read a paper (Nowakowski 1991, 11, 86–93) about a similar settlement in Cornwall and the archaeologist found a line of pits through the centre of the settlement. These were then described as ritual pits (always a good term when one does not know what something is). When one has experienced building such a settlement one knows instantly what the pits are, they are daubing pits. When you are mixing the silt, straw, soil and water together to make the daub for the walls of the houses, it is impossible not to dig a small layer of soil away from the ground with each mix. After a while these daubing pits get too deep to dig the daub out, so you move along the field to find another flat piece of ground and continue the same process. What is left is a line of pits in the middle of the settlement which might well have been used later as fire pits or even for rituals but were originally formed as a by-product of the building process.

We reconstructed an interesting hut from the same Cornish site at Trethellan, years later (Nowakowski 1991, 64–68), which was 6 m in diameter. It was interesting because the ring of post-holes on the outer wall were not close enough together to indicate they were made for a wattle wall. There was also a ring of inner posts the same size 60 cm inside the hut. Also set off on its own, but not in the middle of the hut was another post, which was a bit of a mystery. This was an intriguing hut plan, so it was something we wanted to try. Our first thoughts were that the hut was possibly a winnowing hut. This is somewhere in the dry you could toss the harvested grain up in the air from a tray on a windy day to blow the chaff away. This would leave the grain in the tray clean and ready for processing into flour. The extra post, however, was in the way of the central space, but we put it in place with the other posts this time making them all 2 m high to see if the winnowing hut theory worked. It was not until we set the posts into the ground with the odd inner post too that it became obvious what it was for. It looked like one of the supports for a built-in ladder to get to a possible platform in the roof space. This validated the winnowing shed theory as sacks of grain ready to be processed could have been kept up in the roof space. We cut notches into the inner post and the ring-beam post next to it 20 cm apart to make the rungs of the ladder to climb up into the storage area. We then laid a wooden floor or loft in the roof space and once the hut was finished attempted to winnow grain in it on a windy day, and found it worked very well (Fig. 4.15, a).

Over the years, however, we decided to make a weaving shed out of this hut and added more posts on the outer wall so we could wattle and daub it, but we only made the daubed wall 1 m high, so it made a light and airy hut. It then had a continuous window around the walls making it a perfect workroom for any activity requiring light under cover (Fig. 4.15, b).

There were many examples of the evolution of original houses for other purposes over the years as we built a settlement, just the same as a typical prehistoric settlement would have done. That is why hut plans are sometimes baffling to archaeologists when

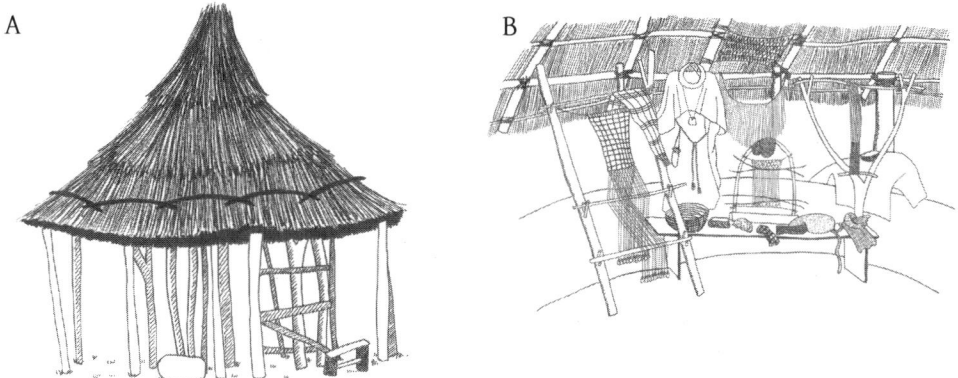

Figure 4.15: a) Hut 141 as a winnowing shed; b) adapted with added walls into a weaving shed

excavating them. People adapt their homes as a new generation might do with an inherited house today. The new occupant might knock a wall down, block a doorway or build an extension just as we found ourselves doing with the roundhouses in our reconstructed settlement. This I feel validates my theory that we as a people have not changed as much as we think we have since prehistoric times. People will always try to make the best out of what they have.

Conclusion

This is my approach to discovering the artisans techniques of prehistoric Europe. In order to even begin to replicate any artefact one has to imagine the possible practical uses of the item and keep this in the forefront of one's mind. Artistic embellishments can be added after the artefact is fulfilling its practical purpose, also it is too easy to assume an item is ritualistic just because it is not obvious what it was used for.

To reiterate:

- Do not acquire any training in the discipline to be replicated, the only way to truly research the subject is to look at the artefact with a clear mind.
- Identify the materials it is made of.
- Assess the fabric of the tools available during the period it was made.
- Be prepared to question an established archaeological theory about an artefact, if it does not seem practical.
- Be prepared to change your assumption about how the artefact was made whilst making it. Let the logical route guide you and keep your plans fluid. As an idea comes into your mind whilst in the middle of a process follow it and see where it takes you. Then through trial and error the creative person is likely to discover a similar route of manufacture that our ancestors did.

A holistic approach to the discipline of replicating prehistoric artefacts is essential I believe. Specialising in one discipline, for instance ceramics or textiles, gives the artisan a blinkered viewpoint on prehistoric daily life. On many occasions I have found it is this multi-disciplinary approach that leads on to conclusions that might not have been acquired if only studying one skill set.

There is a good example of this concept when I was commissioned by the BBC to replicate some of the artefacts associated with a 26,000-year-old body from a cave called Paviland on the south Wales coast. The body was covered in red ochre and wore a loin cloth decorated with winkle shells (*Littorina littorea*) and a fox (*Vulpes vulpes*) tooth pendant on a twine around the neck.

The television company thought it would be good if I cooked some winkle shells as they might have done during the period for the filming and of course make the fox tooth pendant. Making the pendant was a simple task, just drilling a hole in it with a flint awl and threading some twine through it. I cooked the winkles by heating stones in a fire and dropping them into a leather bag which was filled with water and winkles. Once cooked the presenter asked how they might have got the winkles out of the shells without the aid of a metal pin (the traditional method used for this process in Wales).

Suddenly I knew exactly how they did it and it was with the crescent moon shaped fox tooth pendant which was hung around the neck of the body. We tried it and it worked perfectly, much to the disdain of the archaeologist who specialised in the body and had always thought it was a ritualistic pendant associated with the crescent moon. If I had not acquired the knowledge of how to cook the winkles, the question would not have been asked of me and the most likely use for the pendant would not have come to light.

I hope these case studies have inspired readers to arm themselves with my set of basic rules and try and discover for themselves some of the skill sets of our ancestors. Recycling is not a new concept, nothing would have been wasted in a settlement context. Waste would have been utilised or been an inspiration to an artisan of another discipline. I have used these skills to put myself in the mindset of our ancestors for over 25 years and found this multi-disciplinary approach I have developed has led me on to many discoveries.

Over the years I have reached a conclusion that the creative person today is not so dissimilar from our prehistoric forebears as most of us would like to think.

References

Audouze, F. & Buchsenschutz, O. 1992. *Towns, Villages and Countryside of Celtic Europe*. London: Batsford.

Blackmore, C., Braithwaite, M. & Hodder, I. 1979. Social and cultural patterning in the Late Iron Age in southern England. In B.C. Burnham & J. Kingsbury (eds) *Space, Hierarchy and Society. Interdisciplinary Studies in Social Area Analysis*. Oxford: British Archaeological Report S59, 93–111.

Buckley, V. 1990. *Burnt Offerings*. Dublin: Wordwell.

Gibson, A. 2002. *Prehistoric Pottery in Britain and Ireland*. Stroud: Tempus.

Harrod, L. 2000. Letters Page, *British Archaeology Magazine* 51, 24.

Nowakowski, J. 1991. Trethellan Farm, Newquay: Excavation of a lowland Bronze Age Settlement and Iron Age Cemetery. *Cornish Archaeology* 30, 5–242.

Parker Pearson, M. 1990. The production and distribution of Bronze Age pottery in south-west Britain. *Cornish Archaeology* 29, 5–32.

Renfrew, C. 1977. *Alternative Models for Exchange and Spatial Distribution.* New York: Academic Press.

Smith, G.H. 1987. The Lizard Project: Landscape survey. *Cornish Archaeology* 26, 13–68.

Spindler, K. 1994. *The Man in the Ice.* London: Weinfeld and Nicholson.

Wood, J. 1999. Making a Spear and Ice Man's Outfit, *British Archaeology Magazine* 49, 8–11.

Wood, J. 2004. The possible use of fire-cracked stones in ceramic production and recent research on the Otzi grass cloak. In O.V. Smyntyna (ed.) *The Use of Living Space in Prehistory: Papers from a session held at the European Association of Archaeologists sixth annual meeting in Lisbon 2000.* Oxford: British Archaeological Report S1224, 119–121.

Chapter 5

Pathways

Timothy Darvill and Elizabeth Poraj-Wilczynska

Introduction

Walking away from the neatly bounded, regular, categorically defined ontologies of positivism and western science as practised over the past 300 years, past the dung-heap of post-modernist self-referential hyper-critical pluralism, we find ourselves in new fields where thinking is more fluid, more porous, and more emotionally stimulated. Gone are the constraints of Cartesian Dualisms, binary oppositions, and simplistic back-projections of modern life onto other people's existence. In their place is a recognition of dwelling as experience, structures as transformed materials, time as a social phenomenon, and agency as a network of interpenetrating interests. We enter an exciting, novel, and interesting space that Eduardo Viveiros de Castro conceptualises as a kind of cosmological perspectivism in which 'the world is inhabited by different sorts of subjects or persons, human and non-human, which apprehend reality from distinct points of view' (Viveiros de Castro 2015, 195). Here there are no subjects and objects in the traditional sense, no explanations or interpretations, rather a series of intersections between constructed understandings, cosmologies, culture, nature, and the supernatural – networks of existences, events, and controlled equivocation that allows those prepared to make a journey to take any number of pathways through worlds created from their own existence as well as the effects of spaces, materials, environments, and beliefs encountered along the way.

In this chapter we would like to take you, the reader, on five intersecting journeys around the Neolithic long barrow known as Belas Knap, high on the Cotswold Hills of Gloucestershire in the west of England (NGR: SP 0209 2554). Along the way we will explore the shapes, forms, space, textures, colours, feel, and resonances of the long barrow itself, and the responses of human and non-human beings to the place and its changing environment. Our aim is to use a series of works created by one of us (EPW) to question conventional perspectives and use the archaeological imagination to open up new realms of thinking, experience, and understanding.

To the land of Belas Knap

Belas Knap stands at around 300 m above sea-level on high ground overlooking the valley of Beesmore Brook, a west flowing river that joins the Isbourne that, in turn, joins the mighty River Severn, Britain's longest river. The first work, *Barrow Lands* (Fig. 5.1) captures the horizontal and vertical dimensions of this landscape. It shows the barrow on the horizon viewed from the east across the valley, silhouetted against light from the winter sun. The trees have been left out to enhance the stark edge of the scarp. If you stand directly below the barrow, in what is now Humblebee Wood, you are in a dark, dead, cold environment. Even in summer it is cold here. This view helps imagine the journey from the valley bottom into the dark space of the wood and then upwards through the long steep climb up to the barrow on the plateau beyond. Moving out of the dark of the valley into the light airy hilltop is a journey of transformation. Was it part of an initiation? Was it a rite of passage?

Approaching Belas Knap from all other directions is more gentle. The second work, *Chamber* (Fig. 5.2), is an imagined aerial view of the journey from the west towards the barrow whose footprint appears as a shadowy outline slightly off centre. The contours of the land flow under, over, and around the mound. Female shapes are everywhere, folds in the land that somehow harmonise the landscape with the barrowscape. On the coloured version of this work, a slightly dished red triangle is positioned within

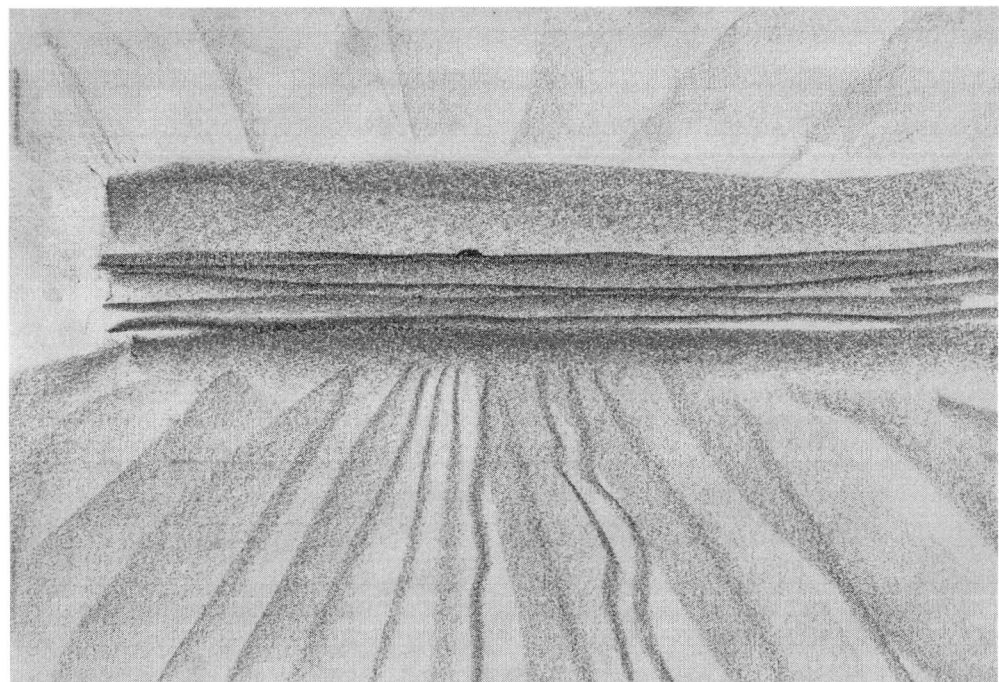

Figure 5.1: Barrow Lands. 2007. Pastel on blotting paper (Elizabeth Poraj-Wilczynska. Copyright reserved)

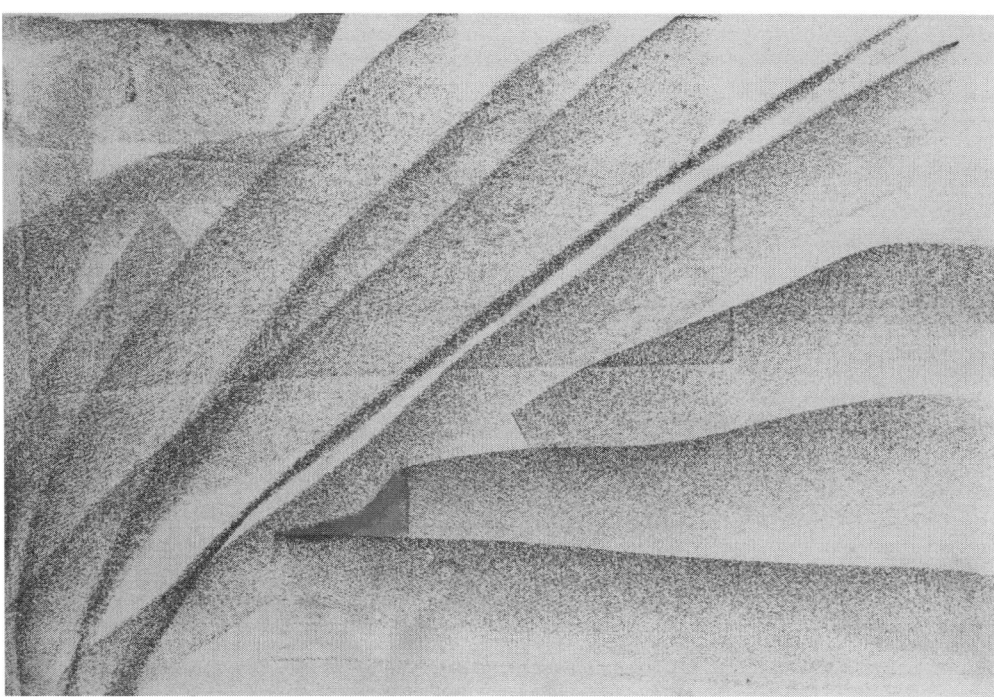

Figure 5.2: Chamber. 2009. Oil pastel on blotting paper (Elizabeth Poraj-Wilczynska, copyright reserved)

the folds. This shape represents reflected light from the north-west chamber as the sun sets Lpposite it on the 21 December: Winter Solstice. This event is vibrant, the warm red of the sun and the deep orange of illuminated Cotswold stone merge as one. If the chamber were opened at this propitious moment in the cycle of the sun it would have created a sense of awe and magic.

Scattered across these landscapes around Belas Knap are fragments of antiquity in the form of complete and broken flint tools and debris from working flint and stone, all of them pieces lost by those that built, used, or knew the long barrow in its prime. Alastair Marshall has systematically walked the fields all around and identified a concentration of worked flints and domestic stonework debris at Humblebee some 300 m to the south (Marshall 1985, 46). But the connections go deeper, as the materials of the land beneath our feet were selected and used to build the barrow; parts of the land are transformed to become elements of the monument.

Today's thin red-brown Cotswold soils cover a rich creamy yellow limestone bedrock. Highly layered and banded, its variety of texture and form is extraordinary. Sea-creatures fossilised millions of years ago in the Jurassic era emerge from the stone and litter the landscape; round echinoids, spiral-shells, and bivalves known locally as 'devil's toenails'. The layers of fine-grained rock contain still smaller fossils and preserve the root patterns of long-dead plants, visible now as holes running through the stone. What did the ancients make of this stone? What did they see in it? Was it living or dead? Active or passive?

The third work, *Stone Worlds: Belas Knap Cosmology* (Fig. 5.3), brings together found elements of the world around Belas Knap that have been re-arranged to resemble a neo-**lithic** cosmos. It represents interactions with the place over the past ten years. All of these objects are part of the 'spirit of the place'. They are all locally distinctive, and occupy certain specific areas of the landscape: Place of the holed-stones; Place of the fossils; and, possibly, Place of the stone ancestors. Barrow and stone lie together in the landscape sharing position, shapes, and textures.

In the centre there is a barrow/axe shaped stone fashioned from a rough lump of limestone found in the west field. This stone was shaped using stone-on-stone techniques; grinding and polishing by hand. The process took two weeks, and the stone was carved on site while EPW slept in the mound and lived by the barrow. Carving stone is deeply absorbing and brought immersion in the landscape and the accompanying soundscape. Animal and bird sounds from the fields and woods, along with the sunrise and moonset, helped to shape the stone. Now they all dwell within it, along with the elements rain and wind, sun and moon, and of course a large part of the maker.

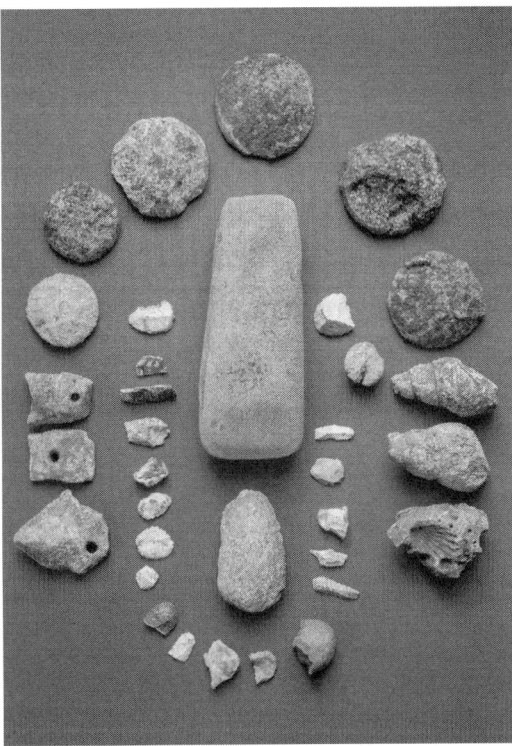

Figure 5.3: Stone Worlds: Belas Knap Cosmology. 2019. Found natural objects, worked flint, hammer stone, carved limestone axe (Elizabeth Poraj-Wilczynska. Copyright reserved)

Echinoid fossils represent the moon's cycle of risings and settings from left to right over the stone barrow/axe. On the left are natural holed-stones – totem stones – easily found on the ground surface in Humblebee Wood. On the right, large fossilized snails and shell imprints representing a band of fossils found when the earth is disturbed by ploughing to the west of Belas Knap. Also a small bivalve, a typical female shape.

The worked flints were found near the fossil band west of the barrow. Placed around the sides and front of the centre-stone they represent the people who built the mound and its associated community. The large fossil bivalve just below the centre-stone represents stone ancestors. There is a fragment of hammerstone within the flint line. This stone, a water worn pebble, was chosen for its strength and probably brought from the stream in the valley below. It was the first of these stones to be collected.

The holed-stones and the fossils' forms all appear within the large stones chosen to construct the mound. In particular the blocking stone (if in its original position) in the north-east chamber has a large and unusual hole in it. The fossil shapes of echinoids can be seen on the back stone of the north-west chamber.

Overall, the Belas Knap cosmology explores the many different lives in stone, and stone ancestors. Were the barrow builders constructing a miniature landscape for the dead to inhabit in spirit form?

Back to the Knap

The curious name is believed to derive from two Old English words: *bel* meaning a beacon; and *cnaepp* meaning a hilltop. But it is far older than the time when these words were commonplace. It was built around 3800 BCE, perhaps as a series of successively larger and more complicated structures, starting as a simple portal dolmen and ending up as a trapezoidal long barrow 50 m long, 18 m wide at the north end decreasing to 10 m at the southern tip, and perhaps 4 m high. There is a deep forecourt defined by projecting horns at the higher and wider end of the mound, opening to the north. Two chambers, each approached by short passages, open from the east side, and one from the west. A fourth chamber may once have existed at the southern end of the mound.

Viewing Belas Knap in the early 1920s, archaeologist O.G.S. Crawford described it as 'one of the most perfect (and instructive) chambered long barrows in Gloucestershire' (Crawford 1925, 67), but what is visible on the site today is the result of recent reconstructions. The earliest recorded excavations were carried out in 1863–64 by L. Winterbotham and continued in 1865 by Joseph Chamberlayne assisted by W.L. Lawrence (Lawrence 1866). This work revealed the poor state of the site and showed that robbing had been going on since later prehistoric times. Despite some attempts at restoration and conservation, co-ordinated by Mrs Emma Dent of Sudeley Castle, the condition of the site deteriorated through the later 19th century and, in 1928, the barrow was transferred to the Guardianship of what was then the Office of Works. This was the Government's department responsible for ancient monuments under the *Ancient Monuments Consolidation and Amendments Act 1913*, a piece of legislation that had transformed approaches to the management and conservation of ancient remains (Thurley 2014). Extensive excavations were undertaken in 1928, directed by W.J. Hemp (Hemp 1929), and in 1929 and 1930 under the direction of Sir James Berry (Berry 1929; 1930). Overlapping the excavations was a programme of restoration work under the supervision of Ralegh Radford who, at the time, was Inspector of Ancient Monuments for Wales. Some of the work was included in the Government's Increased Employment Programme, and the restoration was completed in summer 1931 (Radford 1930). The monument as rebuilt is a very fine example of the 'false-entrance' or laterally-chambered type of Cotswold-Severn long barrow and is widely used as a model of its kind (Darvill 2004, 80).

Other than routine maintenance, little has been done to Belas Knap since it was restored. A guide-leaflet was published in 1966 (Grinsell 1966), variously revised and reprinted in later years, and a succession of interpretative panels placed at the site tell of changing understandings of how long barrows were built and used. Management works have been undertaken in the area to secure access for visitors, as it is a popular place.

In 1966 much of the Cotswold Hills was identified as an AONB – an Area of Outstanding Natural Beauty (Martin *et al.* 1990). Originally conceived as a designation focused on 'natural beauty' the fanciful nature of separating 'natural' from 'cultural' was quickly recognised and later interpretations include concern for wildlife, physiographic features, and cultural heritage as well as the conventional concepts of landscape and scenery (Natural England 2008, 17–19).

From the early 1970s a long-distance footpath known as the Cotswold Way wound its way through 164 km of varied landscapes from Chipping Camden in the north to Bath in the south, passing Belas Knap near the northern end. In May 2007 the status of the footpath was elevated to that of a National Trail (Hayne & Hayne 2009) and the number of users multiplied considerably. The Cotswold Way Association has estimated that the path is used by about 210,000 walkers each year (CWA 2019), to which must be added the hundreds who complete shorter lengths and the many who make the journey to Belas Knap just to visit the barrow. Nowadays it is unusual to find oneself alone at the barrow during daylight hours and, on calendar festivals such as solstices and equinoxes, and Beltane, Samhain, Imbolc, and Lughnasadh, there are ceremonies and the deposition of offerings in the forecourt and around the chambers.

Belas Knap now stands contained and packaged within a walled enclosure separating it out from the woods and fields around about; a wall as a cultural device that artificially disconnects the ancient monument from the ancient landscape. But wait. Look at the barrow. Here too we find a wall defining the edge of the mound, a cultural device used by the original builders to separate the barrow and its content from the wide open spaces around about. What did that barrow originally look like? How was it conceived by those who built and used it? Was the manicured enclosed structure that we see today anything like the monument envisaged in the minds of its builders? Has the spirit of the barrow been contained, or is it free?

The fourth work, *Storm Catcher* (Fig. 5.4), reimagines the barrow viewed from the west. The barrow is stripped of its grassy covering. Bright yellow limestone and large limestone slabs cover the mound and chambers. The barrow is now in a clearing with a fence keeping wild animals out. Fenced pathways lead from all four directions to this place. Narrow straight paths for the living and the dead. The dead in spirit form must be allowed to travel, but only at certain times contained within known areas. On the top of the mound there are wooden posts like totem poles and tall dead trees. These are to attract lightning.

The aim was to explore the idea that Belas Knap might have been built in a location already known as a sacred place. A place where the gods communicated with the living. A place where lightning struck, where the sky echoed, and cracks in the

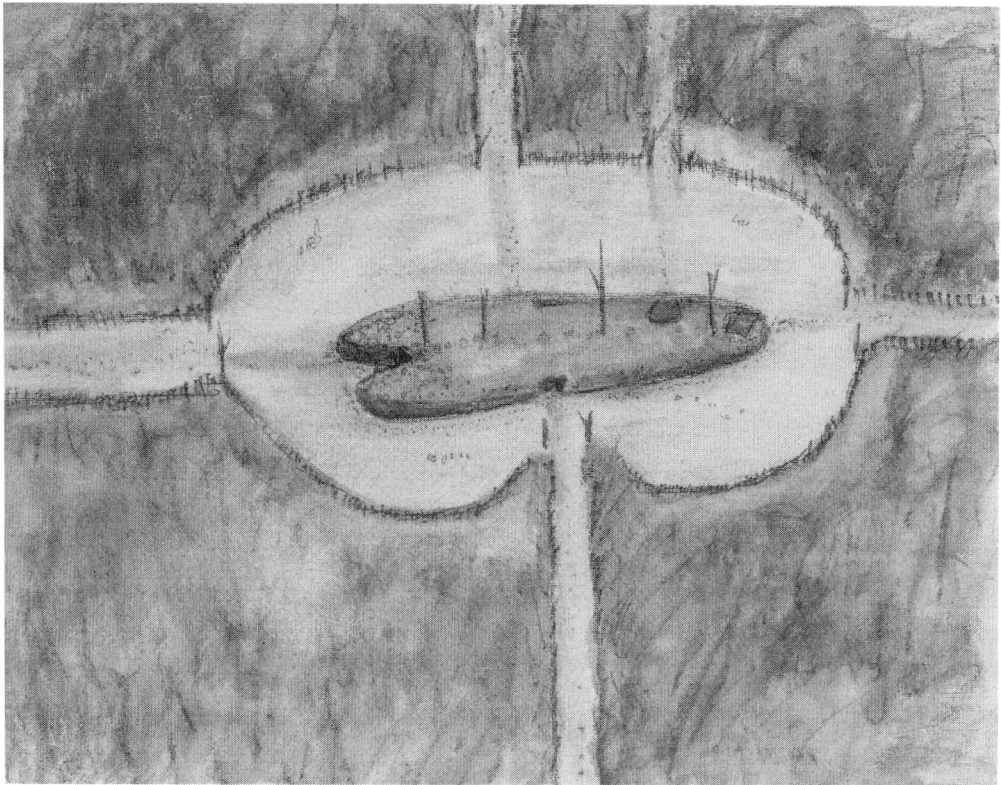

Figure 5.4: Storm Catcher. 2012. Mixed media (Elizabeth Poraj-Wilczynska. Copyright reserved)

cosmos opened up. A place where the living could see the dead travelling to the upper realm on a silver river.

If true, then it would be good to build a mound to keep the ancestors safe until the gods spoke. They would need wooden poles on top of the mound to attract the lightning, mimicking trees in Humblebee Wood that have been scorched and blasted apart by storms in recent times. And certainly Belas Knap has extreme weather in the form of storms that travel up from the valley below. Great clouds billow, suddenly materialising right above the barrow. There are also many fissures channelling underground water below the barrow. A cluster of aquifers lie to the south near the edge of the mound. These too seem to make it a good place for lightning to strike. Could this place have been considered special back into the Mesolithic or beyond?

Amongst animals and people

Animals are listening. They always have been. People too. Every sound changes the dynamic of the landscape, changes the pattern of movement. Observing animal tracks in the snow-covered fields around Belas Knap reveals networks of intersecting pathways like common threads made by wild animals. Foxes, deer, dogs, badgers, rabbits, hares and many smaller creatures criss-cross the land (Graves & Poraj-Wilczynska 2009, 182). Inside the barrow things are different. Animals are few and domesticated. A study by Richard Thomas and Lesley McFadyen (2010, 98) revealed that comparatively few animal remains were found during excavations at Belas Knap, and that cow and pig were the only ancient species identified. The fifth work, *Animal Magic* (Fig. 5.5), builds a map of animal locations and foraging places within the Belas Knap landscape.

The long barrow is in the middle with Humblebee Wood to the left. Animals of the dark wood live here. Wolves, bears, and wild boar. Owls are hidden in the chaotic mass of fallen trees and blackthorn. To the south deer graze and a hare sniffs the air. The moon is rising to the south. To the west a shadow path leads to the winter solstice sun. Stick figures walk here, their long shadows entering the open north-west chamber. A raven rides the breeze and a large eye, the eye of the wind, watches over all. There are snakes, large female adders who have made their nest in the stone quarries. A colony of glow worms lives right by the forecourt.

This picture documents animal interaction with a human place. By spending time with the animals it becomes clear when they are acting out of character, and apparent what it signifies. If the hare is absent from his regular spot there will be rain. If the raven from the west wood is flying unusually high in circles with his mate there will

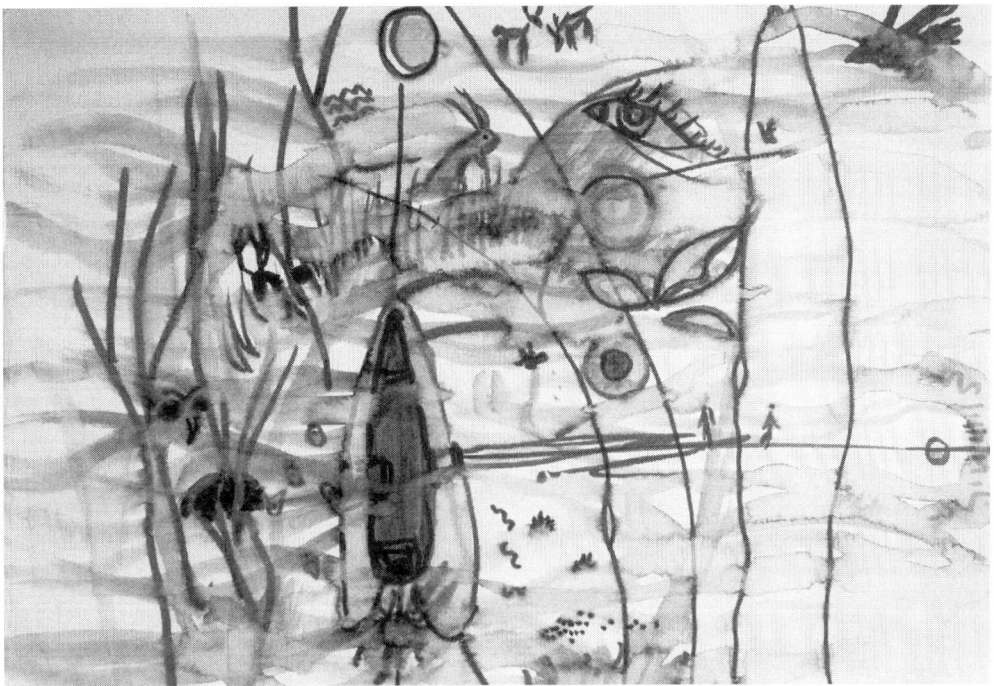

Figure 5.5: Animal Magic. 2018. Watercolour and ink on paper (Elizabeth Poraj-Wilczynska. Copyright reserved)

be high winds by the end of the day. And the animals are constantly talking to each other and listening to the land. Their presence and behaviour dictates the observers' movements; respect their space and they pay little attention to the movements of others ... or so we think.

What of the people Mf the Knap? Modern people mainly move in formal lines using the defined footpaths that countryside stewards have created and marked for them. Structured people in a structured world, only occasionally venturing off course to catch a photograph or pick some berries in the hedge. In the past things might have been different.

Inside the mound things are certainly different. More than 30 individuals are represented amongst disarticulated spreads of bones found in the chambers; men, women, and children who lived during the thirty-eighth and thirty-seventh centuries BCE. All have now been removed. Recent studies suggest that some people, especially males, died violent deaths as they were struck on the head with a stone axe (Parsons 2002; Schulting & Wysocki 2005). At once missing yet present, the people of Belas Knap exert a strong influence on how the place feels.

The human remains from Belas Knap now reside in several different locations. The main collection is in Cheltenham Museum. There are a few small scraps of bone and a single vertebra in Winchcombe Museum. But the bones once held in Cheltenham College have disappeared and no doubt there are other pieces scattered in private collections. It is deeply saddening that the essence of these people is not where it should be – at the barrow. It is not practical for the remains to be returned to the barrow but works of art provide one way of bringing them back together at least in spirit. Many visitors to the barrow remark about the peace and tranquillity they feel standing on the mound so we may assume those who once lay within feel safe and secure.

The sixth work, *Journey of the Mind* (Fig. 5.6), overlays an image of a skull from the barrow onto an image of Humblebee Wood. It aims to explore the sensory experience of the woods, the sky, the outline of the mound, and the people within. It is based on time spent

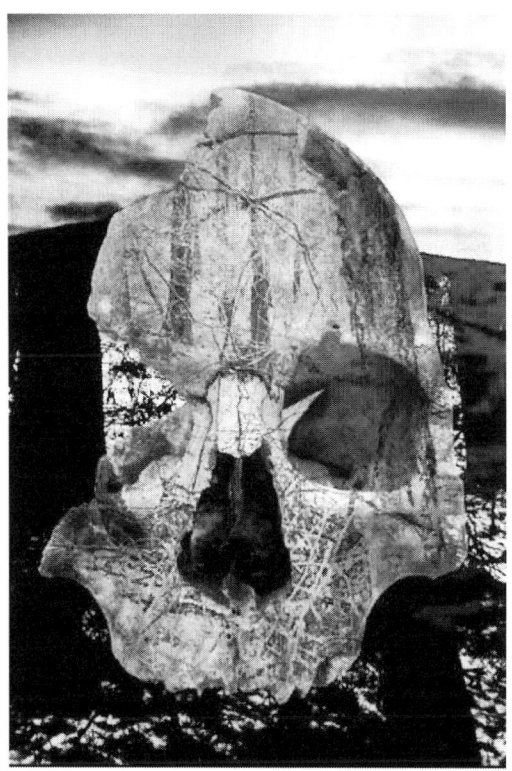

Figure 5.6: Journey of the Mind. 2012. Photomontage (Elizabeth Poraj-Wilczynska. Copyright reserved)

photographing and drawing the skeletons, a hugely humbling experience. They may be perceived as dry shells or husks with no life left inside, but at the same time they feel strong. They have presence, particularly the skull used in this work.

Journey of the Mind is also a piece of music composed of sounds from the mound and its landscape. Stone on stone, a bone tinkMing over the surface of the wall of the forecourt, animal and bird sounds, and more. The compilation of sound culminates in an orchestral piece and takes the listener on a journey from the distant Jurassic period to the present day; it sounds like the land itself.

Looking forward, one of the long-term goals of EPW's project is to make a facial reconstruction of one of the skulls, as the 'face of Belas Knap' would really help the public to engage with the site. Access to the Knap can be tricky for school parties especially, so the face of Belas Knap as part of a small travelling exhibition with the music and images could bring new ways of engagement with the site to new audiences.

Negotiating symbols and meanings

What does the long barrow mean? How might its form communicate with people, animals, stone, trees, deities, and the cosmos? Monuments such as long barrows are often seen as central places or territorial markers, ways of literally engraving the identity of a community onto and into the landscape (Renfrew 1976). But how does the monument itself figure in such schemes? Looking at a long barrow and climbing over it to explore its form and shape conjures many images, but two stand out: similarities between the outline of the mound and the shape of contemporary stone axes, and the resemblances between the overall form of the mound and the female body. The two need not be mutually exclusive; the barrow can be both at the same time or either separately according to context and circumstance. The axe could be seen as representing death and certainly there is evidence that just such implements were used to dispatch people from the world of the living to another place (Schulting & Wysocki 2005). Was that transformation part of a cycle at the other end of which is birth?

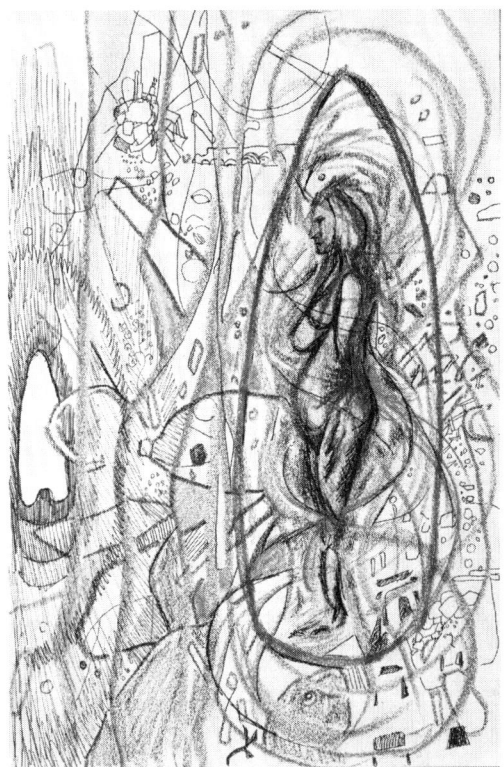

Figure 5.7: Deep Time. 2015. Mixed media (Elizabeth Poraj-Wilczynska. Copyright reserved)

The seventh work, *Deep Time* (Fig. 5.7), explores these interpenetrating meanings across the millennia. It was created on site during the winter of 2015 after spending several days and nights sleeping in the fossil field and having conversations with the mound. At times it felt as if the barrow itself was creating the work and, like many of the works discussed here, it came into being very quickly as if channelled.

The many layers in this image try to evoke a sense of time passing, both geological and archaeological time. The human figure inside the axe outline is a pregnant female. She is opposite the mound looking east. Her body contours, arms, and breasts are made of axe shapes. She is the life-force within the stone. The secret language of the barrow outline wants to reveal itself. The barrow, axe, and human form are one and the same.

The female form transposed onto the barrow is explored further in the eighth work, *Drawninward* (Fig. 5.8). Painted whilst in the forecourt facing the portal, normally a dark shadowy place, this moment was in August as a full moon rose directly above the mound, illuminating the horns of the barrow in a soft, silvery, rounded way. Every contour became gentle and feminine. Could the forecourt have been a place to give birth? The female form overlaying the barrow invites us to enter an inner world. The forecourt area funnels us into the portal, an entrance to the body of the mound. In this image the barrow is female, but at other times it is male and occasionally animal. It is always animate, shape-shifting, evolving, changing as this work tries to capture.

Moonlight, in particular, changes perceptions of the site and provides a view that few people experience today. When approaching the barrow from the north-east the mound feels vast and towering. Its outline is a replica of the horizon, and one wonders whether this was the time that Neolithic people came to the site. As reconstructed, the higher northern end of the mound reflects the trapezoidal shape and seems a logical way of rebuilding a solid structure based on its plan. But this may be too simplistic. Early excavators suggested that the chambers may have been covered in separate mounds later joined together, and taken alongside the presence of many

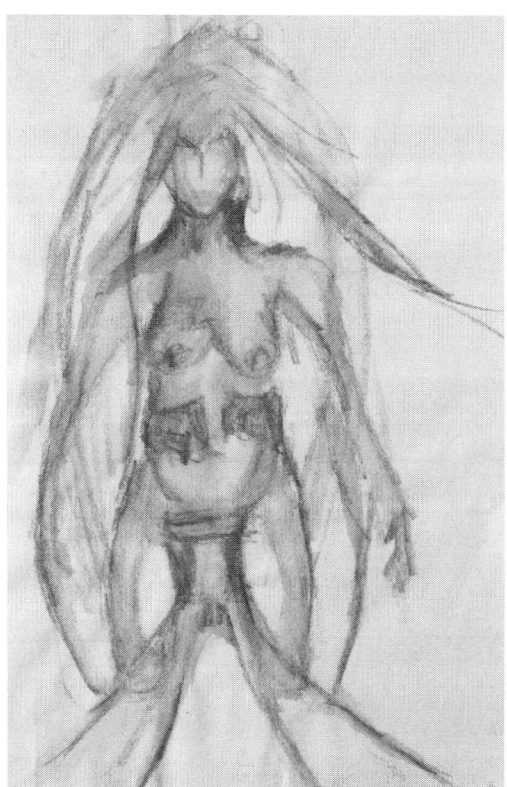

Figure 5.8: Drawninward. 2016. Watercolour (Elizabeth Poraj-Wilczynska. Copyright reserved)

internal walls subdividing the mound into cells or structural units it may be better to think of the mound as an altogether more lumpy structure with higher areas and hollows in its upper surface, more like the form of a human body lying on the landscape.

Nowadays seen as a static monument of constant form, might experiences of Belas Knap have been structured around transitory phenomena? The way that frost lines can be seen across the forecourt floor in winter has been discussed elsewhere (Graves & Poraj-Wilczynska 2009, fig. 11) but snow is another interesting aspect of high altitude monuments in winter, wrapping the site, colouring it white, and providing a plastic building material that can be shaped and moulded. The ninth work, *Ice Maiden* (Fig. 5.9), shows a female figure modelled in ice and snow. She came to life during a period of deep snow and freezing conditions in January 2009. Rising from the top of the barrow the kneeling pregnant woman gazes out towards the winter sunset. She is positioned half-way along the axis of the mound, above the two back-to-back chambers. The axis of the barrow marks the end of the flat lands where the scarp edge drops steeply away. This is a place of transition; light flat open fields to the east, dark steep cold wooded slope to the west. The Ice Maiden turns her back on Humblebee Wood, dark and full of danger. Instead, she faces the setting sun, she is transformed for a short time every day from ice to fire.

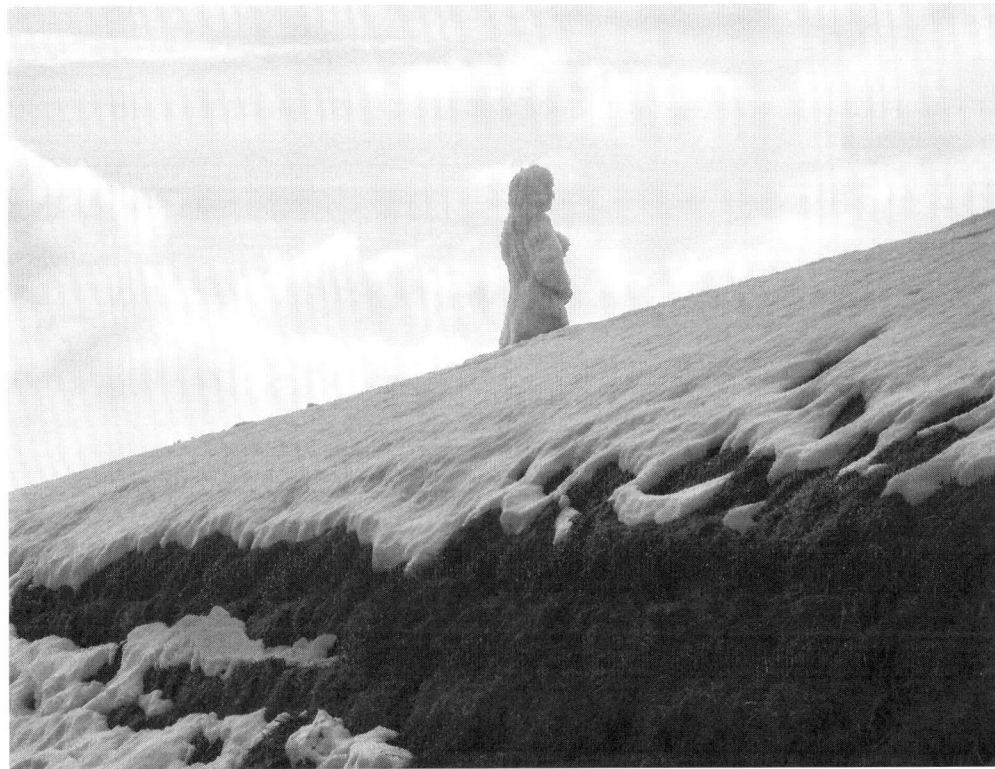

Figure 5.9: Ice Maiden. 2009. Snow (Elizabeth Poraj-Wilczynska. Copyright reserved)

The sculpture was created with help from Jo Webb, a local artist who has lived on the land hereabouts all her life. The journey to the barrow was difficult and felt like a pilgrimage. But instinctively we worked together to create a response to the place even though we had no idea what it would be. There were no discussions about it beforehand, we just knew we wanted to make something from the barrow's snowy cover, and the Ice Maiden took form.

Naturally, the Ice Maiden melted a fraction every day, but she lasted nearly 2 weeks. During this time we documented her demise until she became one with the earth. Like a seed from the sky she found her home in the pregnant body of the barrow. And as the barrow lies beside the Cotswold Way we observed people's reactions to the Ice Maiden. Strangely, very few visitors ventured onto the mound to take a closer look, many perhaps felt uncomfortable glancing up at her so walked smartly away.

Up stairways to heaven

Standing at the barrow and lifting one's eyes skyward the landscape becomes a sky-scape coloured by blues, greys, whites, pinks, yellows and reds during the day, a dark groundmass punctuated by constellations of stars and the Milky Way at night. And revolving through both the great glowing orbs of sun and moon. The unusual north–south orientation of Belas Knap (most long barrows on the Cotswolds are broadly east–west) may be the result of having to accommodate pre-existing structures into the trapezoidal form of the long barrow, or perhaps because of the need to parallel the course of the Beesmore Brook. Either way, there are few obvious celestial movements that could be linked with the overall positioning of the barrow mound or viewable from the forecourt at the north end. A possible exception is the bright star Deneb in the constellation of the Swan (*Cygnus*) that Pamela Armstrong suggests would have been visible setting on the horizon when standing in the forecourt (Armstrong, ongoing PhD research; see also North 1996, 44). More obvious is the orientation of the north-west chamber that opens towards the setting sun, its fading light penetrating most deeply at sunset around the Winter Solstice. The north-eastern chamber has early morning sun for a short time, the light snaking around the bulbous left hand side blocking stone allowing the right-hand side stones and a small portion of the right lower back stone to be illuminated.

Sitting in these chambers, especially at dawn or dusk, is an extraordinary experience and emphasises that the skyscape is as powerful as the landscape in terms of how the world is experienced. And it is not just the sun, moon, and stars. Clouds and weather are important too. Rain, snow, fog, and frost can be felt; they come and go, they change how the world looks, and how it is perceived. They change the sound-scape too. More ethereal, yet every bit as powerful is the transformational power of the wind. The tenth work, *The Southwest Wind* (Fig. 5.10) lifts the perspective from the land to the sky and tries to catch the breeze. The image was created in the wind, and the accompanying project notes, written on site at the time it was done, record something of the experience:

'The southwest wind rushes across the flat lands towards the barrow twisting this way and that urging everything in its path onward. The southwest wind is a singing wind. It swirls and curls around you. If you walk with your back to it you will feel compelled to keep turning around to see who it is whistling a tune right behind you.

This wind will only let me walk its way. I have found shelter in an old quarry to the south of the barrow. I want to draw this wind-person. It is harsh with the smell of ice and stone. When I arrived there was a dense cloud cover, shapes of trees moving in and out of thick grey murk.

The sun is trying to push through but the wind brings dense clouds across the horizon. In the field skylarks are chasing sunspots. In my quarry shelter bare shafts of hazel thrust upwards around me.

Last year's oak leaves make a good bed as I lie down and listen. Far down the slope behind me in Humblebee Wood a thin constant roar is flowing from north to south. Up top I am being buffeted, even in my quarry the west wind is rushing in and over me. It is urgent no time to stop or slow down, on a mission but going where?

I open my eyes to clear blue sky. Silhouetted against it just one oak leaf still attached holding on. I congratulate it, and wonder when it will finally fall to earth to join its brothers and sisters. The north crosswind from Humblebee begins to roar louder, rising to meet the winds flowing from the west. A voice from the dark wood, powerful and forbidding, says time to move.'

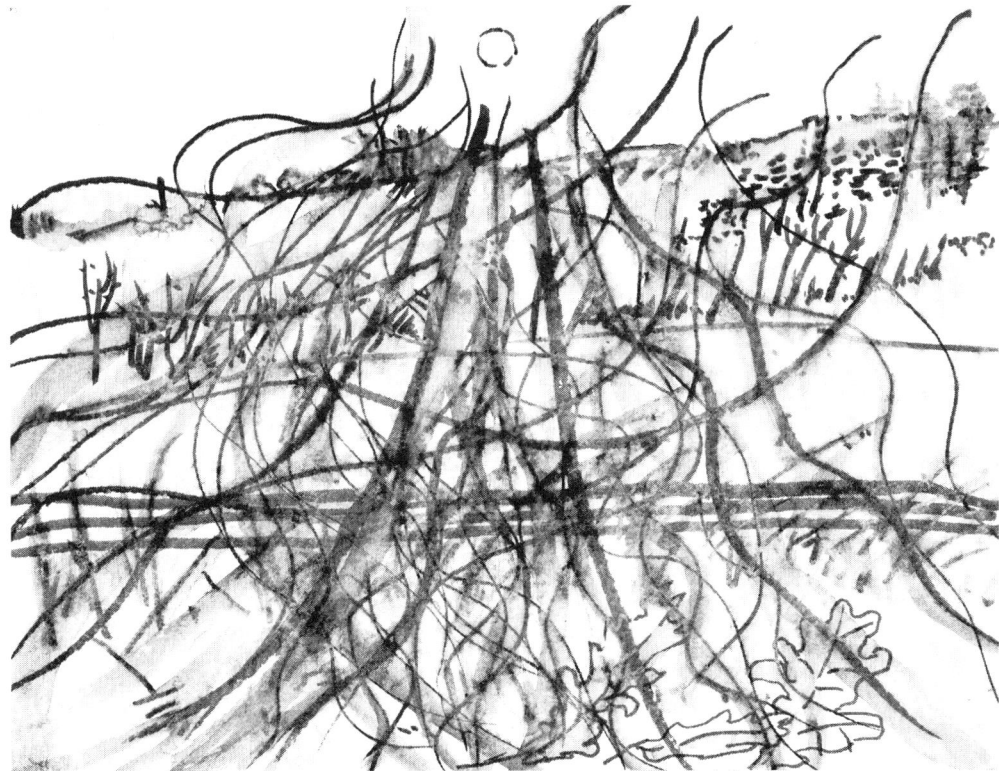

Figure 5.10: The Southwest Wind. 2019. Mixed media (Elizabeth Poraj-Wilczynska. Copyright reserved)

Each wind has a different property and different qualities. Waiting for the wind can take a long time and requires patience. The sound of the wind blowing across the entranceways to the chambers is haunting and loud. Standing on top of the mound in strong wind is difficult and dangerous. But following the winds gives new perspectives and new questions. Where, for example, would you place your settlement when the prevailing wind is from the west? Will it carry your scent towards the animals you want to hunt when you set out of from camp each day? And with reference to the barrow, might the smells of decomposing bodies best be carried away from living areas? It is important to understand the wind, and what it might bring to your doorstep, and it is certainly a subject for further investigation.

Discussion

Through the ten works explored here, new approaches open up to a well-known long barrow in the Cotswolds. Each work highlights bonds between past and present. In a sense the barrow itself becomes a time-machine that enables us to interpret it, providing us with pathways that link the hopes, fears, and beliefs of its builders and users with our present-day imaginations. Through these works, based upon sensory and immersive experiences, we can speculate about those far-off times and raise questions and concerns.

Embedded art practices further our understanding of prehistory, and importantly, allow us to learn about our connections to landscape and environments past and present. In this way the inherent creativity allows ideas and themes to be explored in ways that would not otherwise be possible because the work itself condenses out ways of thinking, ways of seeing, such that the viewer is challenged to figure it out and perhaps arrive to new or different understandings (cf. Renfrew 2003). Controlled equivocation in the archaeological imagination allows journeys along intersecting networks of thought-paths, endlessly changing track, looping back, exploring the edges, or resting at comfortable nodes. It enlarges our domains of knowledge beyond descriptions of material culture and social behaviour to consider agency and emotion. Within a perspectivist context it is not just the human mind that has to be considered, what about the social lives of animals, trees, rocks, and so many other elements and dimensions of the world that all inhabit. These too can be drawn into the constructed images of art-based archaeology.

Acknowledgements

We would very much like to thank Dragoş Gheorghiu for inviting us to contribute to this volume an essay about our favourite site and one that has occupied both of us for many years; Jo Webb for help with the Ice Maiden; Pamela Armstrong for guidance on the skyscapes of Belas Knap; and Alistair Marshall for conversations about flint scatters and cosmology.

References

Berry, J. 1929. Belas Knap long barrow, Gloucestershire. Report of the excavations of 1929. *Transactions of the Bristol and Gloucestershire Archaeological Society* 51, 273–303.

Berry, J. 1930. Belas Knap long barrow, Gloucestershire. Second report: The excavations of 1930. *Transactions of the Bristol and Gloucestershire Archaeological Society* 52, 123–150.

Crawford, O.G.S. 1925. *The long barrows of the Cotswolds*. Gloucester: John Bellows.

CWA (Cotswold Way Association), 2019. The Cotswold Way. [Organization website available at: http://cotswoldwayassociation.org.uk/our-other-trails/. Viewed: 28 Sept 2019]

Darvill, T. 2004. *Long barrows of the Cotswolds and surrounding areas*. Stroud: Tempus/History Press.

Graves, T. & Poraj-Wilczynska, E. 2009. 'Spirit of place' as process: Archaeography, dowsing and perceptual mapping at Belas Knap. *Time and Mind* 2(2), 167–194.

Grinsell, L.V. 1966. *Belas Knap*. London: Department of the Environment.

Hayne, T. & Hayne, B. 2009. *Cotswold Way*. Hindhead: Trailblazer.

Hemp, W.J. 1929. Belas Knap long barrow, Gloucestershire. *Transactions of the Bristol and Gloucestershire Archaeological Society* 51, 261–272.

Lawrence, W.L. 1866. An account of the examination of a chambered long barrow in Gloucestershire. *Proceedings of the Society of Antiquaries of London* (2nd series) 3, 275–80.

Marshall, A. 1985. Neolithic and earlier Bronze Age settlement in the northern Cotswolds: A preliminary outline based on the distribution of surface scatter and funerary areas. *Transactions of the Bristol and Gloucestershire Archaeological Society* 103, 23–54.

Martin, J., Plowden, M., Nicholson, D., Howard, P. & Hamilton, P. 1990. *The Cotswold Landscape: A landscape assessment*. Cheltenham: Countryside Commission (CCP 294)

Natural England, 2008. *State of the Natural Environment 2008*. Peterborough: Natural England (NE85). [Available online at: http://publications.naturalengland.org.uk/publication/31043?category=118044. Accessed 27:09:2019]

North, J. 1996. *Stonehenge. Neolithic Man and the Cosmos*. London: Harper-Collins.

Parsons, J. 2002. Great sites: Belas Knap. *British Archaeology* 63, 18–23.

Radford, C.A.R. 1930. Belas Knap long barrow. *Transactions of the Bristol and Gloucestershire Archaeological Society* 52, 295–299.

Renfrew, C. 1976. Megaliths, territories and populations. In S.J. DeLaet (ed.), *Acculturation and Continuity in Atlantic Europe*. Bruges: DeTempel, 198–220.

Renfrew, C. 2003. *Figuring it Out. The Parallel Visions of Artists and Archaeologists*. London: Thames and Hudson.

Schulting, R.J. & Wysocki, M. 2005. 'In this chambered tumulus were cleft skulls ...': An assessment of the evidence for cranial trauma in the British Neolithic. *Proceedings of the Prehistoric Society* 71, 107–138.

Thomas, R. & McFadyen, L. 2010. Animals and Cotswold-Severn long barrows: A re-examination. *Proceedings of the Prehistoric Society* 76, 95–113.

Thurley, S. 2014. *Men from the Ministry. How Britain Saved its Heritage*. New Haven CO: Yale University Press.

Viveiros de Castro, E. 2015. *The Relative Native. Essays on Indigenous Conceptual Worlds*. Chicago IL: Hau Books.

Chapter 6

Art in the corporal memory and in the mental imagery

Dragoş Gheorghiu

Introduction: efficiency as art

Over time, as an experimentalist, I was often challenged when having to differentiate between archaeology and art. For me, since the 1980s, artistic research has functioned as a kind of metaphorical archaeology, being situated in a border area, that I would call art-chaeology (Gheorghiu 2009a; 2009b; 2009c).

It is from this art-chaeological perspective that I will present my research from the last decade, similar to an art diary, in which the first person textual narrative is supplemented with different visual examples of multiple archaeology experiments. I believe it is thus possible to present the experience of the subjective personal approach in an intelligible form, and to show the cognitive analogies between the archaeological research and the artistic practices (see Gheorghiu & Barth 2019).

By studying the natural forms I became acquainted with the concept of efficiency and discovered that part of traditional aesthetic is based, though not explicitly, on this concept. Based on the principle of efficiency, the brain organises and simplifies the composing elements of complex images, to cite the Gestalt law of proximity (Zakia 1997, 3 ff): 'producing patterns easy to image and remember'. Thus, for example, the pattern of a yarn, or of the wrapping of two threads around a volume, can be perceived as a series of chevrons and lozenges (Fig, 6.1, a, b). The visual pattern thus obtained is the result of efficient gestural ergonomics on the part of the operator and, as the art of traditional societies demonstrates, produces geometric images with aesthetic value. If we look at art 'as a form of acting' (Gell 1998, 67; see also Jones & Cochrane 2018, 20), the aforementioned pattern is one of the simplest artistic productions and a diagram of an efficient and rhythmic body movement.

The embodied mind: body, rhythm, memory and mental imagery

The pattern of lozenges discussed is a 'pattern of embodied experience' (Johnson 1987, 175), respectively of an embodied cognition (Varela *et al.* 1993, 147), which stores

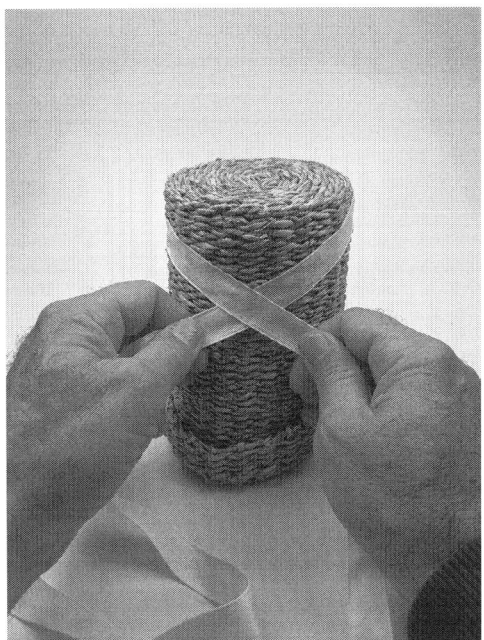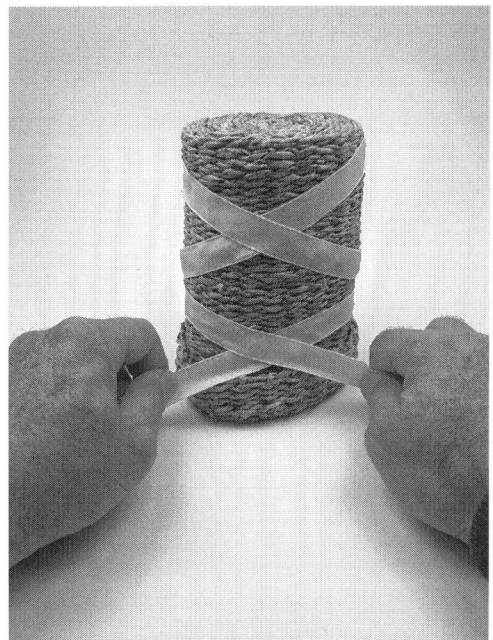

Figure 6.1: Plaiting two stripes around a volume (photo M. Moţăianu)

these aesthetic images in memory. It is apparent that the body memory retains the efficient gestures, just as the iconic memory (Humphreys & Bruce 1995, 192) stores them in imagery.

In the case of rhythmic gestures, the rhythm functions as a fixer of the gesture and of the image in memory. In 'Memory and rhythms', in the second part of the volume *Gesture and Speech* (1993) Leroi Gourhan highlighted the mnemonic character of repetitive gestures in technologies (see also Ingold 2013, 115).

Recent research has revealed the close connection between bodily experience and visual memory which is the result of an embodied mind (Gallaher & Zahavi 2012, 146 ff) 'that shows how knowledge depends on being in the world, a world that is inseparable from our bodies, our language, and our social history – in short, from our embodiment' (Varela *et al.* 1993, 149). Performative actions belong equally to the field of experimental archaeology research, and to contemporary visual and digital art (see Dixon 2007).

Case study 1: the Cucuteni-Tripolye figurines: construction and decoration

In the case of personal experiments, an example of the patterns created by the ergonomically efficient gestures, transposed into imagery through body memory, is that of the Chalcolithic decorated figurines in the Cucuteni-Tripolye tradition (see Monah 2016). A large number of ceramic figurines from the Pre-Cucuteni and Cucuteni A–B cultural phases have engraved lozenge patterns, and in the final period of the

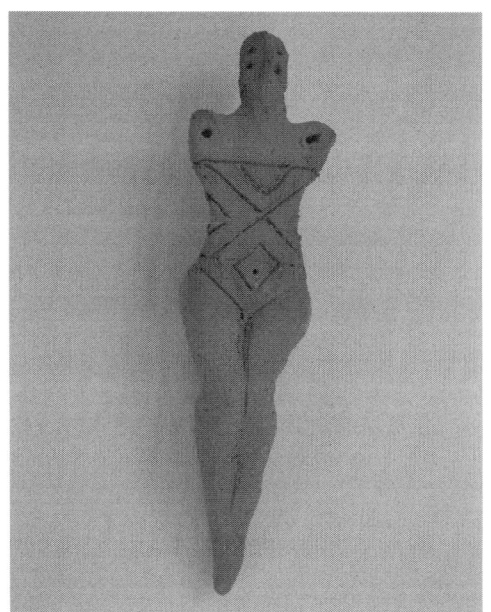
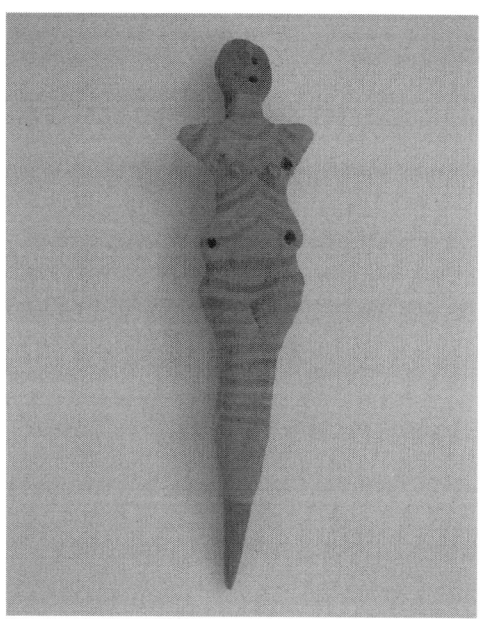

Figure 6.2: Cucuteni figurines; replicas, student work, Vădastra, 2001 (photo D. Gheorghiu)

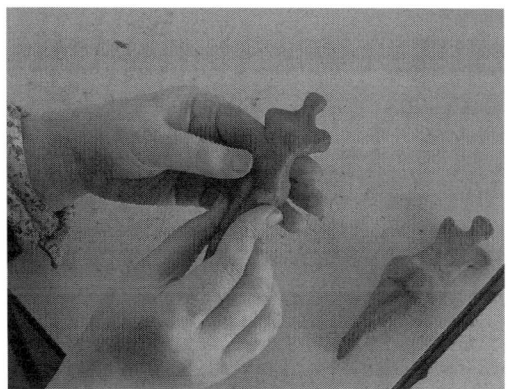

Figure 6.3: The modelling of a figurine, Macao experiments, 2018 (photo D. Gheorghiu)

tradition these patterns are painted in black (Figs 6.2, a, b).

Due to the special body shape, the Cucuteni-Tripolye figurines were initially identified as representing female bodies (Gimbutas 1992), despite the fact that the large size of the pelvis is the result of the special modelling technology utilising equal clay modules (Gheorghiu 2005). The construction of the figurines is based on a very precise structuring of the modelling gestures (which can be regarded as a technological ritual), which allows for the easy learning of their construction and a memorisation of the operative chains after a long time. With only three spherical clay modules available, the performer can achieve an anthropomorphic form after a small number of hand movements, by shaping simplified anatomical parts. The experiments we have carried out in the last decades with children and adults have shown the ease of memorizing the construction process. Even after a single attempt, the shape of the figurine and the *chaîne opératoire* stages were stored in the mind of the experimenter (Fig. 6.3).

Decorating figurines with lozenge or chevron patterns can also be easily reproduced and memorized. Previous experiences with braids or braiding, stored in iconic

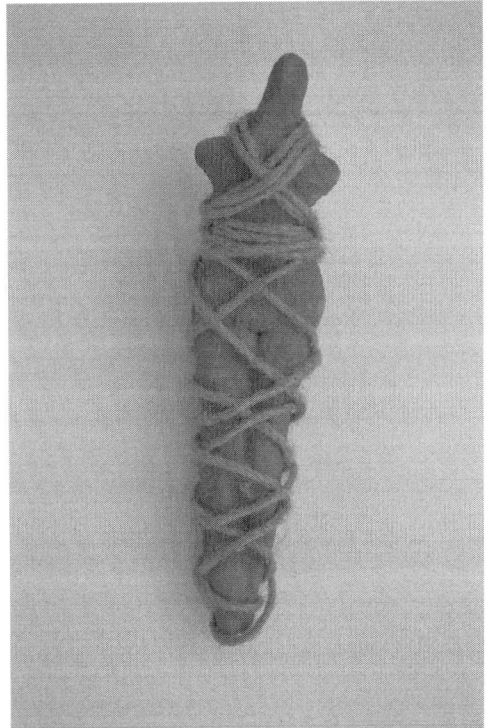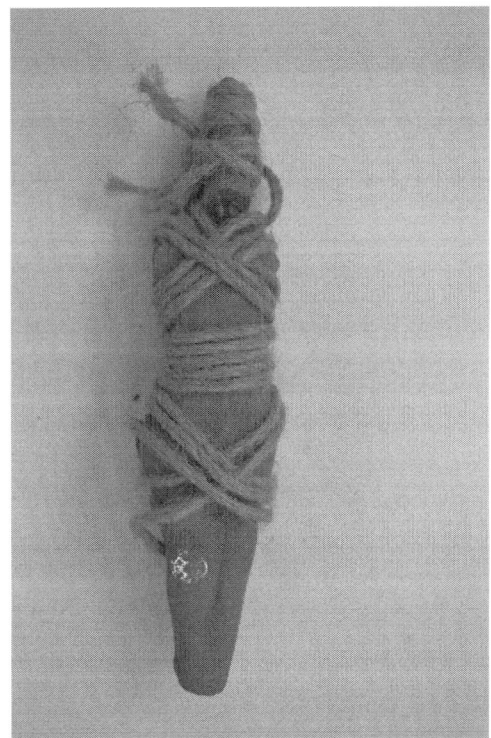

Figure 6.4: Wrapped Cucuteni figurine, following the incised lines on its body (photo D. Gheorghiu)

and bodily memory, have revealed these patterns as representing diagrams of gestures of wrapping figurines (and possibly human bodies) with textile fibres, as experiments have shown (Gheorghiu 1992; 1996; 2003; 2005; Fig. 6.4). It can be said that these patterns on the figurines could be an index of some mummy-like burial customs of the deceased (Gheorghiu 2001).

Case study 2: Barclodiad y Gawres: the engraved stone and the land-art

Another example of how bodily memory resulting from rhythmic activity shapes imagery is that of the engraved stone inside the Late Neolithic passage tomb of Barclodiad y Gawres on Anglesey, Wales (Nash 2008; Nash & Stanford 2011). The tomb, dated to about 3500 BCE, is the interior of a cairn positioned on a high cliff. In 2014, within the GestART project, coordinated by Professor George Nash, I had the opportunity to create a contemporary artwork that would present the monument to the public. While searching for a representative image to achieve this, I identified on the engraved standing stone located to the right of the entrance to the funeral hall, a lozenge pattern similar to the one on the Cucuteni-Tripolye figurines. The incisions were performed in a single episode, without superimpositions. By perceiving the upright posture of the stone as an anthropomorphic feature (Straus 1966, 138; Gallagher & Zahavi 2012, 150), the image mentally overlapped that of the Cucuteni-Tripolye figurines, or of a human body, and I recalled the gestures of wrapping with two textile threads. We repeated these gestures on the stone, using textile bands and strings, in order to embody the ergonomics and the rhythm of the plaiting described by the engraving (Fig. 6.5, a–c).

In this fashion I tried to embody the Neolithic pattern carved on the standing stone and then sought an artistic way to reveal this experience to the world. In order to create a great degree of visibility for the artwork with which we intended to promote the monument, we chose to transform the lozenge pattern into a land-art covering the entire exterior surface of the monument.

In this case the land-art works as an enhancer of the significance of the place (Gheorghiu 2012), able to evoke different elements of the material or spiritual culture of the past. Augmenting the real is a new type of contemporary art (Gheorghiu & Ştefan 2013) that is currently applied in digital art (Geroimenko 2018). Used as an art-chaeological tool, the augmentation can reveal unseen architectural traces or ritual paths and guide the discussion on issues of orientation and positioning of the builders of the past.

The enlargement on a monumental scale of the engraved pattern led to a disembodied experience in a first stage of the work. Thus, a visualisation of the site from above was necessary, which is an imaginary process specific to the projective thinking, then the positioning on the site of the drawing enlarged at scale, followed by the positioning of the landmarks for drawing the lines of the image with the help of textile strips (Fig. 6.6).

This design operation represented a transfer of the initial corporal memory and imagery of the plaiting of the standing stone into a large-scale configuration, or in other words, represented a transfer from an *egocentric* space ('of the perceiving and acting body') in the Merleau-Ponty (1962) sense, to an *allocentric* space ('purely objective, that can be defined in terms of latitude and longitude') (Gallagher & Zahavi 2012, 160). The *allocentric* space, which confronts the mind with unverified bodily situations, represents a first step in the process of splitting which occurs between the body and mind.

The disembodied mind: the post-phenomenological experience of the digital art

Due to the large dimensions of the land-art, a drone was used to shoot and photograph all the artwork from altitude. The ascension of the drone was followed on the display, observing the gradual decrease in size of the landscape and of the artwork (Fig. 6.7, a–d). The visual sensation produced was that of a flight in which the mind was released from the body, a sensation analogous to that of an 'out-of-body experience in which there is a perception of the self and personal identity separated from the physical body' (Winkelman 2015, 331). Such an experience is characteristic of 'shamanic soul flight'. The experience of 'the separation of the soul from the body

Dragoş Gheorghiu

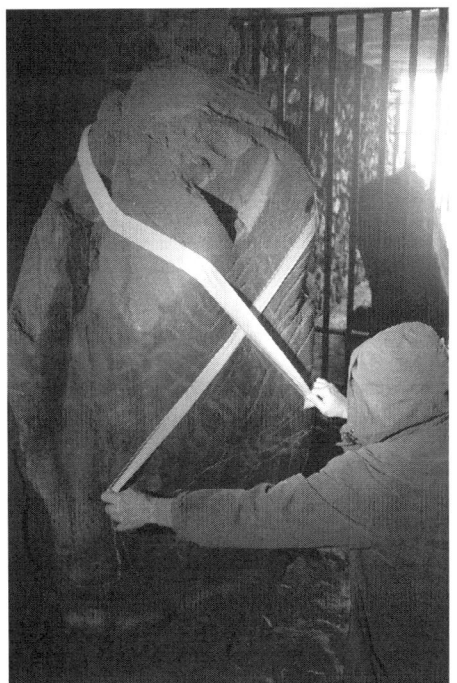
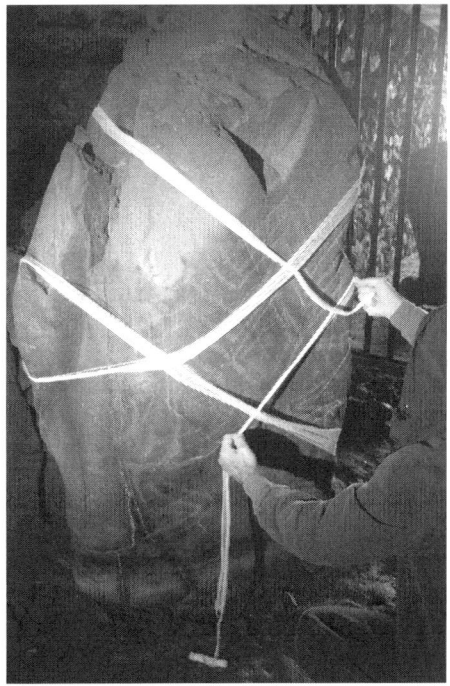

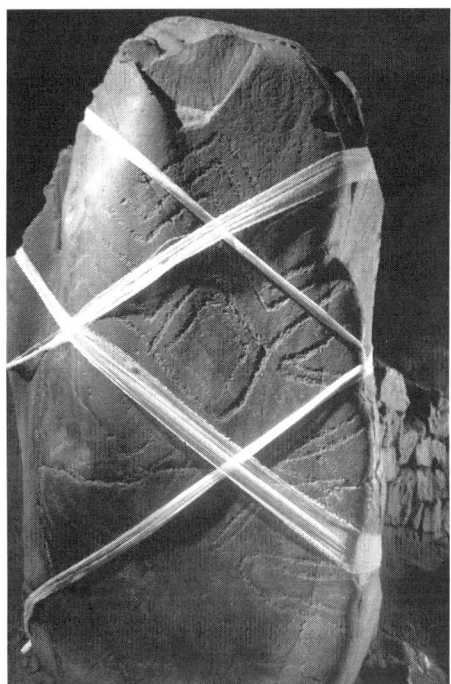

Figure 6.5: Wrapping the standing stone inside the Neolithic tomb, Barclodiad y Gawres, Wales (photo G. Nash)

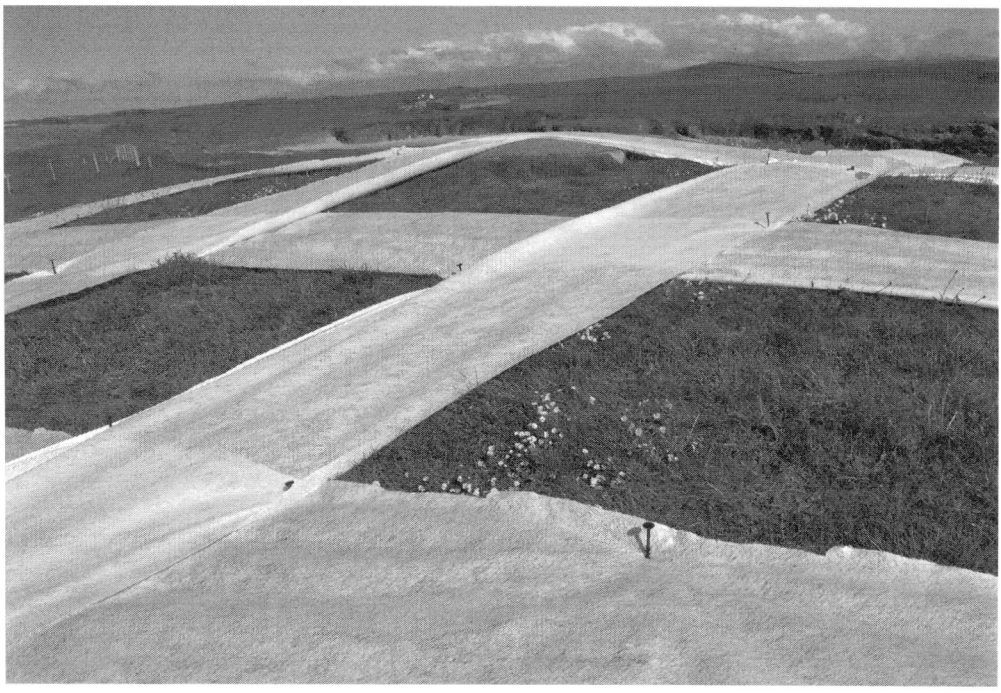

Figure 6.6: The positioning of textile strips on the surface of the cairn, Barclodiad y Gawres, Wales (photo D. Gheorghiu)

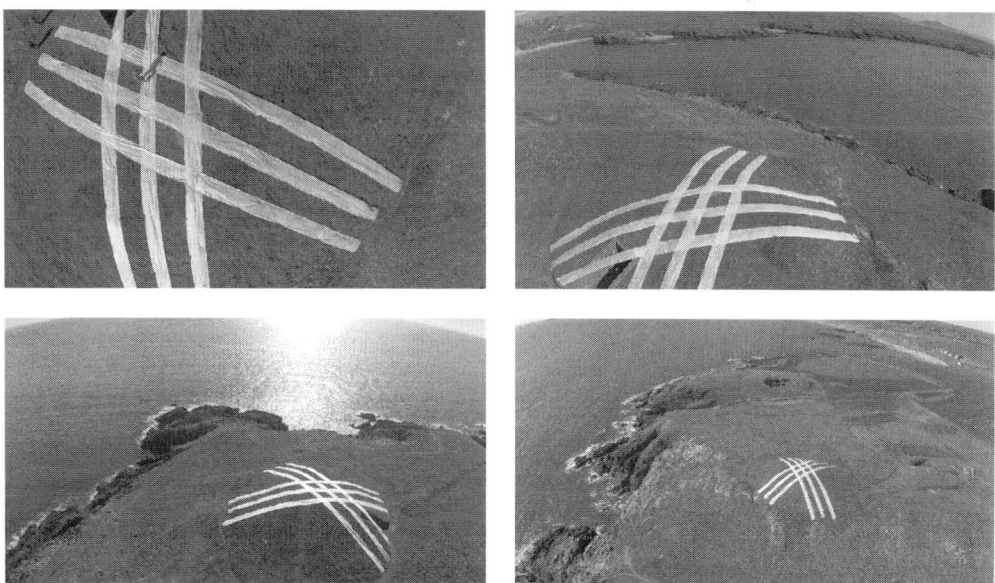

Figure 6.7: The land-art seen from the drone's eye, Barclodiad y Gawres, Wales (photos: A. Beardsley)

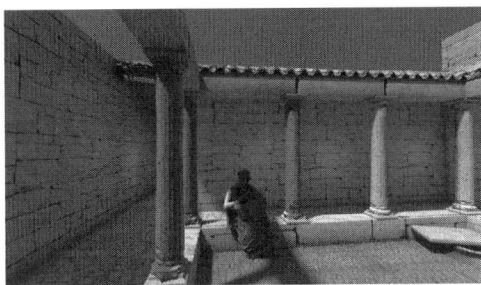 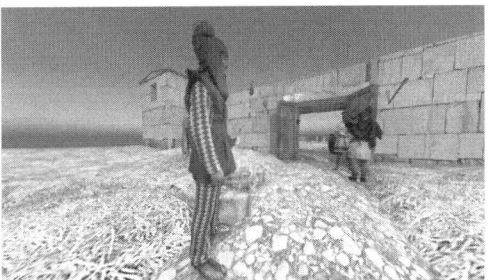

Figure 6.8: Characters scanned in 3D: a) http://timemaps.net/timemap/mangalia/?page_id=2790;
b) http://timemaps.net/timemap/albesti/?page_id=2927 (3D reconstruction by M. Hodea)

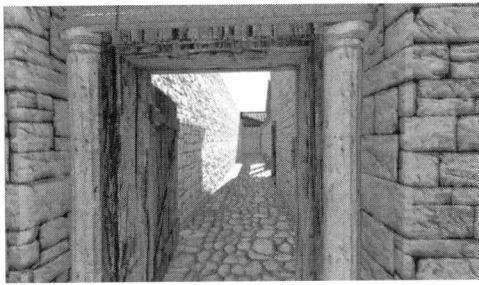

Figure 6.9: Reconstruction of a Hellenistic house, http://timemaps.net/timemap/manga-lia/?page_id=2790 (M. Hodea)

is a human cultural universal and key to understanding a variety of aspects of shamanism' (Winkelman 2015, 332); it resembles post-phenomenological experiences (Idhe 2009; Crystal 2018) of immersion in the virtual worlds. Post-phenomenology deals with this new type of relations between the mind and constructed reality which currently tends towards hyper-realism (Shanks & Webmoor 2013, 88) in virtual reconstructions. Immersions in the virtual worlds produce mental states similar to those of immersion in theatrical works (Reaney 1999), and in this respect an experiment designed to increase the realism of the reconstructions of the past was to insert into them real characters, scanned in 3D and thus transformed into digital objects (Gheorghiu & Ştefan 2015; 2018; Fig. 6.8, a, b).

Another experiment aimed at improving the immersion in the virtual worlds due to corporal memory was realized by increasing the virtuality of the digital artworks; thus, prehistoric and historical reconstructions of context were realised in Virtual Reality, in which were augmented those features of the environment that could stimulate a reaction of the real-world body memory, respectively the illumination, the colours and the textures of objects (Gheorghiu 2015; Fig. 6.9).

Although the visual end result may sometimes seem exaggerated compared to reality, for the person immersed in the virtual reconstruction the spatial experience, although disembodied, was more intense than that experienced in the same space realised in Second Life/OpenSim (Gheorghiu & Ştefan 2019).

The goal of the experiments in real and virtual context reconstructions was not only the reconstruction of architectural structures or objects, but also the study of the spatial experience of the performances, respectively the problems of orientation in space and of visual and kinetic ergonomics. For example, the lighting

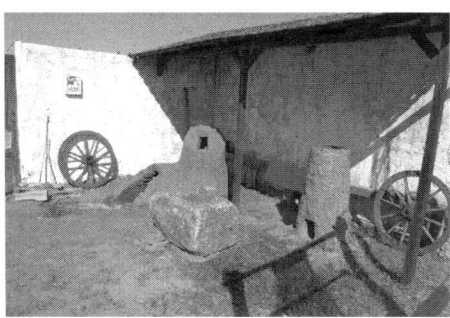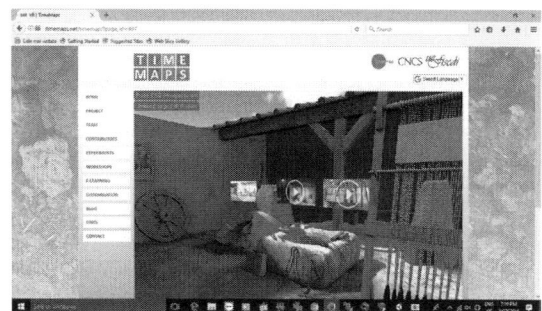

Figure 6.10: a) A Roman workshop reconstructed in Vădastra village (photo D. Gheorghiu); b) a Roman workshop reconstructed in Unity (Golem Studio)

studies performed in the physical reconstruction of a workshop of a Roman rustic villa (Gheorghiu 2018; Fig. 6.10, a, b) were used to increase the degree of realism in the reconstruction of the same construction in Unity (http://timemaps.net/timemap/?PAGE_ID=44).

Conclusions

In the experimentation of the past we combined in the personal experiments both the phenomenological and the post-phenomenological approaches, mixing the experience and the corporal memory with the disembodied experiences, in order to generate an increased experiential perspective of the Past.

Art has been present starting with the generation of basic design and performance patterns and up until site augmentations, or augmented virtualities in virtual worlds. For this purpose we have used different artistic ways of augmenting the real or the virtual to enrich our imagery of the Past.

Experimenting with different traditional materials and textures created a bodily memory that helped me in different cases of pattern recognition or of understanding of a virtual space, for example. In the case of experiments in virtual worlds we have attempted to prevent the disembodied experiences from taking place in *allocentric* spaces, trying through different augmentations to introduce into them the body memory from the egocentric spaces of the experiments in the real world.

Art allowed a transfer between the real and the virtual worlds in the theatrical attempts to reconstruct the 'real' of the past, by inserting objects and real characters scanned in 3D. Apart from the aforementioned, art, like technology, has allowed the production of mental states analogous to shamanic experiences of disembodiment, but also of sensory experiences that IT cannot yet reproduce.

It can be concluded that art, through its many forms and affordances, is an instrument for increasing the capacities of science and that it is and will be present in the archaeological imagination.

Acknowledgements

The experiments described were possible due to a PN II IDEI grant, *Time Maps. Real Communities, Virtual Worlds, Experimented Pasts* (www.timemaps.net). The author thanks all the persons that were involved in the experiments. Special thanks to Dr Livia Ştefan, PhD student Marius Hodea, Professor George Nash, Andy Beardsley, Cornelia Cătuna and M. Bogdan Căpruciu. Last, but not least, my gratitude goes to Dr Julie Gardiner for her long patience during the elaboration of this chapter.

References

Crystal, D. 2018. Postphenomenology and archaeology: Towards a temporal methodology. *Time and Mind* 11(3), 297–304.

Dixon, S. 2007. *Digital Performance. A History of New Media in Theater, Dance, Performance Art, and Installation.* Cambridge MA & London: MIT Press.

Gallaher, S. & Zahavi, D. 2012. *The Phenomenological Mind.* London and New York: Routledge.

Gell, A. 1998. *Art and Agency. An Anthropological Theory.* Oxford: Clarendon Press.

Geroimenko, V. (ed.) 2018. *Augmented Reality Art.* Cham: Springer Series on Cultural Computing (2nd edn).

Gheorghiu, D. 1992. Design in Eneolithic Societies. Unpublished PhD Thesis. Faculty of History, University of Bucharest.

Gheorghiu, D. 1996. Pots and messages: The complex advertising of Eneolithic ceramics. In D.A. Meyer, P.C. Dawson, & D.T. Hanna (eds) *Debating Complexity - Proceedings of the 26th Annual Chacmool Conference, Calgary, Alberta.* Calgary: Alta.

Gheorghiu, D. 2001. The cult of ancestors in East European Chalcolithic: A holographic approach. In P. Biehl & F. Berthemes (eds) *The Archaeology of Cult*, Budapest: Archaeolingua, 73–88.

Gheorghiu, D. 2003. Invisible presence and clay ancestors: Human body and statuettes in Chalcolithic funerary ritual. *Archaeological Review from Cambridge* 16, 143–149.

Gheorghiu, D. 2005. The controlled fragmentation of anthropomorphic figurines. In G. Dumitroaia, J. Chapman, O. Weller, C. Preoteasa, R. Munteanu, D. Nicola & D. Monah (eds) *Cucuteni 120 Years of Research. Time to Sum Up.* Piatra Neamţ: Editura 'Constntin Matasă', 137–144.

Gheorghiu, D. 2009a. A study of art-chaeology, *Archeologia Africana - Saggi occasionali 2005-2009* 11–15, 45–50.

Gheorghiu, D. 2009b. De l'objet à l'espace: Une expérience art-chéologique de la préhistoire. *Etudes Balkaniques; Cahiers Pierre Belon* 15, 211–224.

Gheorghiu, D. 2009c. *Artchaeology. A Sensorial Approach to the Past.* Bucharest: UNArte.

Gheorghiu, D. 2012. Metaphors and allegories as augmented reality. The use of art to evoke material and immaterial subjects. In I-M. Danielson, F. Fahlander, & Y. Sjostrand (eds) *Encountering Imagery. Materialities, Perceptions, Relations.* Stockholm: Stockholm Studies in Archaeology 57, 23–32.

Gheorghiu, D. 2015. Sensing the past: The sensorial experience in experiential archaeology by augmenting the perception of materiality. In J.R. Pellini, A. Zarankin, & M.A. Salerno (eds) *Coming to Senses. Topics in Sensory Archaeology.* Newcastle upon Tyne: Cambridge Scholars Publishing, 119–140.

Gheorghiu, D. 2018. Lighting in reconstructed contexts: Experiential archaeology with pyrotechnologies. In C. Papadopoulos & H. Moyes (eds) *The Oxford Handbook of Light in Archaeology.* Oxford: Oxford University Press [DOI: 10.1093/oxfordhb/9780198788218.013.28]

Gheorghiu, D. & Barth, T. (eds) 2019. *Artistic Practices and Archaeological Research.* Oxford: Archaeopress.

Gheorghiu, D. & Ştefan, L. 2013. In between: Experiencing liminality. *Leonardo Electronic Almanac LEA* 19(1) [https://www.leoalmanac.org/vol19-no1-not-here-not-there/]

Gheorghiu, D. & Ştefan, L. 2015. Augmenting immersion: The implementation of the real world in virtual reality. In W. Börner & S. Uhlirz (eds) *The 20th International Conference on Cultural Heritage and New Technologies CHNT 2015*. Vienna: Museen der Stadt Wien – Stadtarchäeologie, 1–10.

Gheorghiu, D. & Ştefan, L. 2018. A fractal augmentation of the archaeological record: The Time Maps Project. In V. Geroimenko (ed.) *Augmented Reality Art*. Cham: Springer Series on Cultural Computing (2nd edn), 297–314.

Gheorghiu, D. & Ştefan, L. 2019. Virtual art in learning and teaching archaeology: An intermedia to augment the content of virtual spaces and the quality of immersion. In D. Gheorghiu & T. Barth (eds) *Artistic Practices and Archaeological Research*. Oxford: Archaeopress, 166–183.

Gimbutas, M. 1992. *Civilization of the Goddess. The World of Old Europe*. San Francisco CA: Harper Collins.

Humphreys, G.W. & Bruce, V. 1995. *Visual Cognition. Computational, Experimental and Neuropsychological Perspectives*. Hove & London: Lawrence Erlbaum Associates.

Idhe, D. 2009. *Postphenomenology and Technoscience*. Albany NY: Sunny Press.

Ingold, T. 2013. *Making. Anthropology, Archaeology, Art and Architecture*. London and New York: Routledge.

Johnson, M. 1987. *The Body in Mind: The bodily basis of imagination, reason and meaning*. Chicago IL: Chicago University Press.

Jones, A.M. & Cochrane, A. 2018. *The Archaeology of Art. Materials, Practices, Affects*. London & New York: Routledge.

Leroi Gourhan, A. 1993. *Gesture and Speech*. Cambridge MA: MIT Press.

Merleau-Ponty, M. 1962. *Phenomenology of Perception*, London: Routledge & Kegan Paul.

Monah, D. 2016. *Anthropomorphic Representations in Cucuteni-Tripolye Culture*. Oxford: Archaeopress.

Nash, G. 2008. The symbolic use of fire: The case of its use in the Late Neolithic passage grave tradition in Wales. *Time and Mind* 1, 143–158.

Nash, G. & Stanford, A. 2011. Searching for an ancient past, *Minerva* 22(2), 17–21.

Reaney, M. 1999. Virtual reality and the theatre: Immersion in virtual worlds, *Digital Creativity* 10(3), 183–188.

Shanks, M. & T. Webmoor 2013. A political economy of visual media in archaeology. In S. Bonde & S. Houston (eds) *Representing the Past. Archaeology through Text and Image*. Oxford: Oxbow Books, 85–110.

Straus, E. 1966. *Phenomenological Psychology*. New York: Basic Books.

Varela, F.J., Thompson, E., & Rosch, E. 1993. *The Embodied Mind. Cognitive Science and Human Experience*. Cambridge MA & London: MIT Press.

Winkelman, M.J. 2015. Shaman/shamanism. In R.A. Segal & K. von Stuckard (eds) *Vocabulary for the Study of Religion*. Leiden & Boston MA: Brill, 331–339.

Zakia, R.D. 1997. *Perception and Imagining*. Boston & Oxford: Focal Press.

Chapter 7

Modernity and landscape through art: deconstructing the mindset of British contemporary artist James Lawrence Isherwood

George Nash

Introduction

Before I begin to interrogate the concept of art as a social device, I should consider the idea of what modernity is. My subject matter – the mindset of the 20th-century artist James Isherwood (Iddon 2013) took me beyond the artist *per se* to areas of research which I had previously considered to be unrelated to the subject in hand. Clearly there are a number of philosophical and sociological channels that feed in and out the paintings of Isherwood which provide an essential contextual framework to this 20th-century artist, namely social class and politics. Modernity as an analytical concept is closely associated with the ethos of modernism, and relevant to this chapter, modern art. Much of the subject matter incorporated into Isherwood's northern-style paintings show evidence of Marxist principles, revealing something of the man behind the brush (Isherwood came from humble beginnings and hovered on the breadline during his career as an artist). Isherwood, along with his contemporaries was a product (or victim) of the social relations associated with capitalism and shifts in attitudes flowing the succession of war and political change within post-war Britain. Sociologist Michel Foucault (1977) summarised modernity as:

- A historical category that was marked by developments such as a questioning of rejection of tradition; prioritising of individualism, freedom and formal equality;
- Faith in inevitable social, scientific and technological progress, rationalisation and professionalisation; a movement from feudalism towards capitalism and the market economy, industrialisation, urbanisation and secularism; and
- The development of the nation state, representative democracy and public education.

Pertinent to this chapter, I would add to Foucault's general list:

• From imperialism, to nationalism and the fascist state, to liberalism and a European ideology to Neoliberalism and Popularism; and
• Social strangulation to social manipulation, to social awareness.

It is within each of the categories that Isherwood and his contemporaries would have experienced and been influenced by various socio-political ideologies, hence the individual style expressed by Isherwood and, say, his northern-style. Isherwood's years as an artist falls within the 'modern art' period (*c.* 1860–1970) whose beginnings included Baudelaire's works.[1] Like Isherwood, Charles Baudelaire was expressing his views of how art was influenced by the surroundings of the artist; with Baudelaire's it was his experience of living in the *urban metropolis*. Baudelaire was concerned with how to capture and characterise his experiences onto the canvas and how to understand the relationship between artist and subject matter.

Being Northern

The prolific British expressionist and impressionist artist James Lawrence Isherwood was active in terms of painting landscapes from the 1950s until shortly before his death in 1989. Most of Isherwood's formative life was in Lancashire, northern England and it was within the various *scapes* of this part of England where he drew most of his inspiration, creating the so-called 'Wigan Style'. Much of his inspiration came from his mentor L.S. Lowry who himself was a painter influenced by his local working-class surroundings. Lowry once remarked about Isherwood as being 'the man most likely to follow my footsteps'. However, all through Isherwood's working life he was plagued with self-doubt and anxiety.

During his career Isherwood painted over 3000 paintings and for the most part he kept rigidly to an adopted style including many open landscapes and *urbanscapes* which are now considered to be his best-known works. The *politic* of Isherwood's work is clearly reflected, that of the working-class landscape; the deprivation, the grimness and poverty associated with the industrial north. Throughout his career, Isherwood contextualised these landscapes with an occasional expedition to the more salubrious areas of England such as Oxford and London.

Isherwood: a man in a hurry

James Lawrence Isherwood was born in 1917 into a working-class family in Wigan, Lancashire (Fig. 7.1). Isherwood's family were cobblers and he worked in the shop until his father's death in the 1950s. Throughout his life and until his death in 1989, Isherwood had limited commercial success. His paintings were considered by art critics as controversial and sometimes *gimmicky*. His portrayals of the celebrities of the day did much to damage his reputation as a serious artist. Despite this, however, he did become

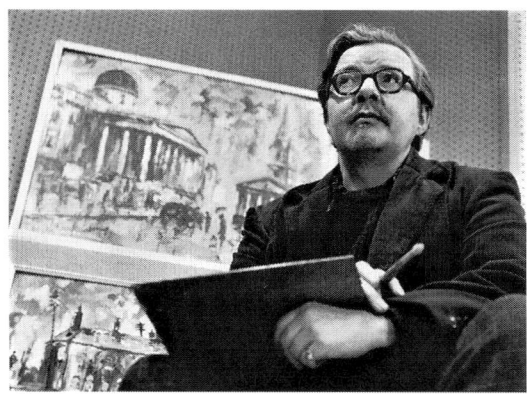

Figure 7.1: Isherwood the artist, taken in the late 1960s. Courtesy of Richard Heath at Isherwood Art (www.isherwood.co.uk)

a respected protégé of British figurative artist L.S. Lowry. Much of the figurative work by Isherwood such as his 'Wigan Women' concept had stylistic associations with Lowry's 'matchstick men' which features in many of his Northern British industrial *scapes*. The Wigan Women concept used blacks and grey hues to portray the bleakness and cruelness of working-class Britain (Fig. 7.6). Paradoxically though, he also painted in vibrant colours similar themes as if to celebrate his working-class roots (Fig. 7.4). Throughout his career he was known locally as 'the eccentric artist', and many of his works were painted under the influence of alcohol. This detail alone may partially account for his sometimes distorted perspective and vibrant colours used on some of his works.

Although much of his work sold during his lifetime, Isherwood's work was not commercially successful until after his death, when his sister-in-law Molly Isherwood and former agent organised a number of exhibitions. Following his death from cancer in 1989 many hundreds of his paintings remained unsold. Even though the man and his paintings were not recognised before his death they are now widely collected, giving Isherwood the respect he should have gained when living.

The archaeology of painting

I suppose it is impossible to ignore the similarities between archaeology and the physical act of painting a picture. Archaeologists, in both methodological and practical ways, are attempting to make a valid interpretation of the past (however deep and distant) by peeling back the layers in order to construct an archaeological narrative. True, there are other means of understanding, say, chronology, typology, and material culture development by looking at a wider context. The complexity of the archaeology, combining all the available evidence, creates a meta-narrative – stories within stories. Using archaeological terminology, stories within stories is represented by stratigraphy: a layer-upon-layer construction of the past. Although successive cultural periods are different from each other there are, albeit tenuous, threads that bind archaeological narration together; usually the physicality of the *site*. The site draws together different groups of people from different times to settle and utilise the same space. Clear evidence for this is within the *urbanscape*. Excavations undertaken by myself during the 1990s in central London, for example, revealed many layers of archaeology – from the Neolithic to the present day, resulting in over 6000 years of occupation from one site. These archaeological subplots can be further investigated to reveal, sometimes, a complex entanglement of human presence (Hodder 2012).

Likewise, the painter behaves in a similar way, not as a chronicler who, through research, would peel away the various histories in order to establish, say, a chronological sequence to reveal cultural development, but more a history-maker, adding the various narratives within a composition to a multifaceted image that will be interpreted in different ways by different people. Take, for example Flemish Renaissance artist Pieter Bruegel's painting *Landscape with the Fall of Icarus* which was painted sometime during the late 1550s (Fig. 7.2).[2] The painting features Icarus, a hero from Greek mythology who is mentioned by the Roman poet Ovid in his *Metamorphoses*. The unsigned painting was discovered in 1912 and recalls pictorially the fall of Icarus after flying too close to the sun. The wings, made by his father Daedalus, were feathers secured by beeswax. In ignoring his father's warning of not to fly too close to the sun the beeswax melted and Icarus fell back to earth, into the sea and drowned. Bruegel's painting shows just the legs, a partially submerged hand and feathers of Icarus, who is kicking at the surface of the water, next to a merchant ship (Fig. 7.3).

The focus of the painting should be on Icarus but, instead, Bruegel places greater emphasis on a ploughman (in the foreground), a shepherd and a fisherman; all three figures appear to not be interested in the calamity that is unfolding. Based on later critical discussions and poetry based on this particular painting, the hidden message within it appears to be the 'ignorance of people to fellow men's suffering' (Kilinski II 2004).[3] Similarly, the poem *Musée des Beaux Arts* by W.H. Auden intimates that the painting depicts humankind's indifference to [godly] suffering by highlighting the mundane events which occur around the death of Icarus. This Renaissance view of an ancient myth with its probable simple message is wrapped up in a complex set of visual narratives that can be deconstructed to reveal a series of layers and processes that extend beyond the original myth. I suppose in terms of physical deconstruction of the painting itself, one could interpret the composition by peeling-off the layers of paint, creating a type of 'Harris matrix' for the mechanics of the painting. What I mean by Harris matrix is the stratigraphic layering of the narratives (rather than the

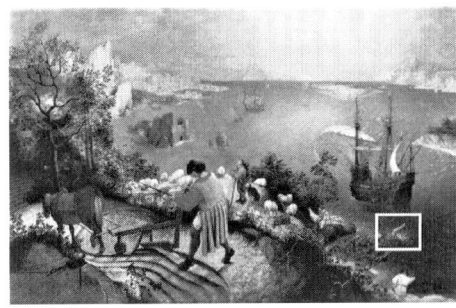

Figure 7.2: Pieter Bruegel's Landscape with the Fall of Icarus *(c. 1558-1560) © Royal Museums of Fine Arts of Belgium, Brussels*

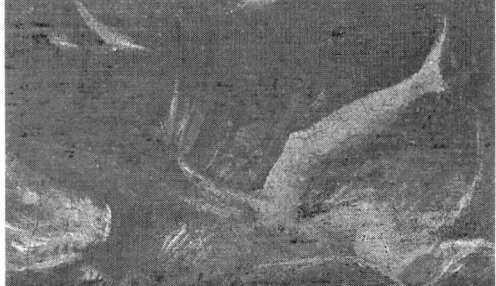

Figure 7.3: Painting detail showing Icarus © Royal Museums of Fine Arts of Belgium, Brussels

physicality of paint) that occurs within a painting; a technique also used exclusively by James Isherwood where stories are interwoven into stories to create subplots (e.g., see Figs 7.4 & 7.5). The stratigraphic make-up of a painting provides the onlooker with a number of subplots that are either associated or not. Metaphorically, there are a number of subplots or narratives ongoing in this painting with the main character portrayed as an insignificant event within a complex setting. The artist appears to be deliberately ignoring the 'main event' and concentrating on the mundane banal scenes that arguably dominate the overall picture narrative.

Similarly, the detail incorporated into Isherwood's *Impression from a Train: Johnson's Cleaning Factory*,[4] 400 years after Bruegel's *Landscape with the Fall of Icarus* conjures a novel image of a view of a northern [English] factory scene but from a train. This scene, although static as a painting, is a representation of a moving snap-shot image from a moving train carriage (Fig. 7.4). The principal narrative of the painting is the Johnson's factory; a scene not unfamiliar to Isherwood's roots and politics. Isherwood's factory (wherever it may be) was a metaphor for working-class ideals and replicates principles of L.S. Lowry's underlying concepts. Lowry, a contemporary of Isherwood, knew of Isherwood's work and was suitably impressed by many of his paintings that portrayed post-war working-class life. Both artists were 'northerners' and both painted *factoryscapes* that focused upon the factory buildings and the employees who worked in them. The scenes usually involved the workers outside the factory gates milling around the streets; reminiscent scenes of workers captured on many early 20th-century films by Michell and Kenyon.[5]

The meta-narrative incorporated into Isherwood's *Impression from a Train: Johnson's Cleaning Factory* involves the factory buildings (the backdrop) and its employees (the principal focus). The factory buildings, by the time this scene was painted, were already relics of a bygone age and would have been demolished within several decades. The Johnson's Factory, probably in Wigan, portrays the classic vernacular of a northern mill, probably mid- to late 19th century in date where many towns of northern Britain (in particular, Lancashire) prospered from the production of textiles. Many skylines of towns such as Bolton, Blackburn and Wigan were dominated by the tall chimneys and the pollution they spewed from their coal-fired machinery. The building provides the setting and the various emotive terms that one would associate with a working-class town – dirt, grime, grubby, smog and stained. These words certainly reflect the principal focus – the people who occupy the foreground

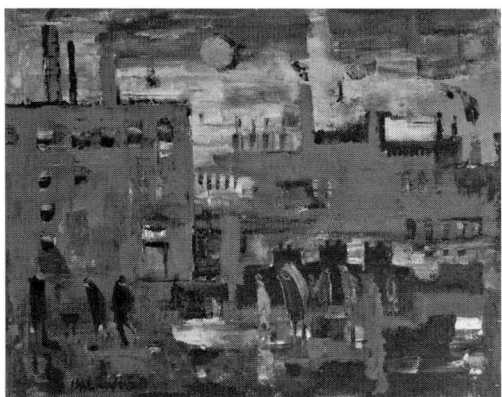

Figure 7.4: Isherwood's Impression from a Train: Johnson's Cleaning Factory *(Courtesy of the Pilkington Collection)*

of the painting. Despite the sunny disposition, most of the figures are painted using dark tones and simple brush strokes, reflecting the general mood of the narrative. However, Isherwood, not without a sense of humour, paints a figure within the left hand side of the picture who is looking down at his right leg. A dog stands close to this figure and has clearly left its excrement on the pavement for this individual to step in (Fig. 7.5). The subplot is subtle and therefore requires the audience to look carefully at the painting in order to fully understand that given the deprivation of such a scene, there is always humour to offset the 'humdrum' of working-class life. Interestingly, several of figures within the painting are similar in style to Isherwood's *Wigan Women* (Fig. 7.6), a unique and unmistakable style that was adopted during the 1950s.

Using the stylistic features from *Impression from a Train: Johnson's Cleaning Factory*, Isherwood's *Wigan Women* applies a similar visual formulaic with a factory backdrop (albeit on the horizon) and human figures in the foreground and 'acting' as the principal focus of the painting. Not forgetting his working-class background, Isherwood uses a watercolour wash to create a simple garment for each of the four main figures. The garment is a [woollen] shawl, a simple wrap-around blanket that would have covered the body and the head. Although this garment would have been woven in many colours, Isherwood chose to paint his figures in blacks and greys, probably representing the hardships of working-class women living in northern England at this time. Using Isherwood's humour for such a dark place, one of the women has her stockings (tights) halfway down her legs to the knees. This trait was commonplace in the working classes because they were used way past their best!

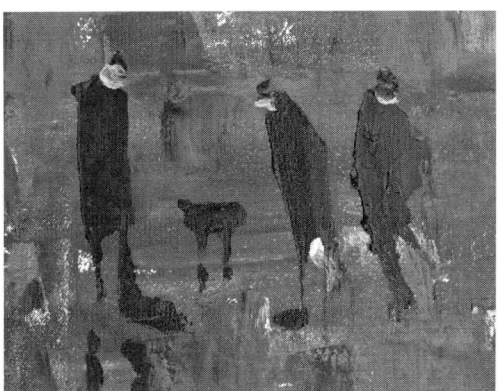

Figure 7.5: An amusing sub-plot involving a dog, its excrement and the unfortunate individual who stepped in it!

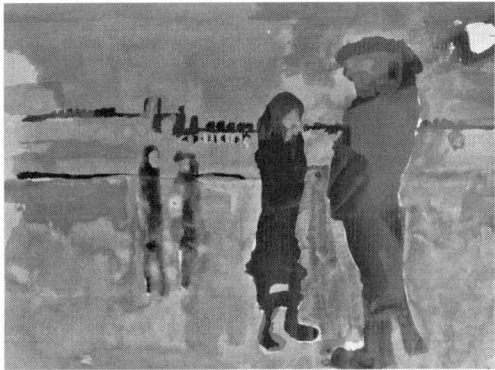

Figure 7.6: Isherwood's Wigan Women *[watercolour] (Courtesy of the Pilkington Collection)*

In complete contrast to the paintings that typify his roots and inheritance, Isherwood also turns his attentions to landscapes that extend beyond the curtilage of northern Britain. Over a 30-year period, Isherwood travels the length and breadth of the country chronicling many notable *scapes*, such as Magdalen College, Oxford (Fig. 7.7). Although the

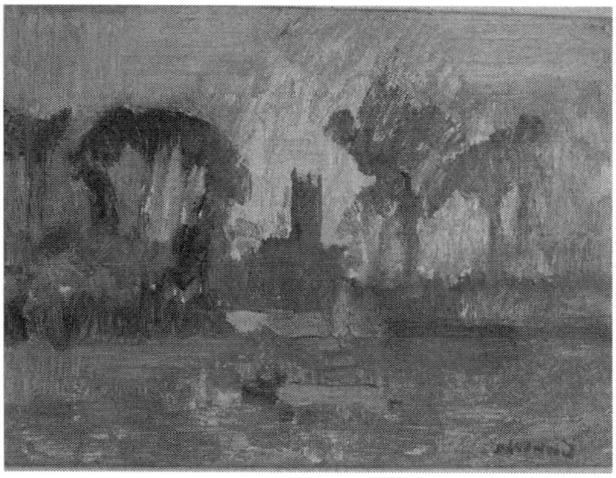

Figure 7.7: Isherwood's Magdalen College, Oxford
(*Courtesy of the Pilkington Collection*)

style employed for both *factoryscapes* and landscapes is similar (and recognisable as an Isherwood style), there are clear differences between the two. With his northern-style, Isherwood veneers the garnish of a class-ridden society throughout, whilst his British landscapes take on a more effervescent sheen, becoming almost apolitical. Here, Isherwood makes full use of his colour pallet to express the vibrancy of the principal focus, such as Magdalen College and its surrounding vista. This painting, along with many others, can be considered almost expressionistic in style, similar to the landscape works of Paul Cezanne and James Whistler. Both these artists and Isherwood made full use of the palette knife rather than the brush, creating a textured and layered surface (see Fig. 7.8).

How did Isherwood, a relatively unknown artist, who during his life craved notoriety, get recognised as one of Britain's most influential expressionist artists? I have suggested earlier that much of Isherwood's subject matter focused on the working class which in many respects represents his early life in Wigan. Isherwood's most productive years were between the 1950s and early 1980s. This period of time witnessed a number of significant changes in artistic style, many of which formed part of a new movement that reflected the optimism of post-war Britain but at the same time animated the plight of a class-ridden society. Here, art becomes a political tool, along with film, literature and performance. Isherwood's *Impression from a Train: Johnson's Cleaning Factory* and *Wigan Women* show the harsh realities of working-class Britain and reflected in the so-named *Kitchen sink* realism, a British cultural movement that developed during the 1950s and 1960s with films such as *A Taste of Honey* (1958) and *Saturday Night, Sunday Morning* (1961). These and many other films applied a style known as *social realism* which portrayed the disillusionment of angry young men and women within society. Like Isherwood, the playwright John Osborne (*Don't Look Back in Anger* [1956]) and Jeremy Sandford (*Cathy Come Home* [1966]) were gritty reminders that Britain had an underclass and raised many issues concerning class, gender, race and sexual orientation. These issues later became the staple diet for the Soap Opera of the mid-to-latter part of the 20th century and beyond (e.g. Tony Warren's *Coronation Street* [1960] and Julia Smith and Tony Holland's *Eastenders* [1985]).

Uncovering the archaeologies of expressionism

The painting style of James Isherwood is firmly within the British expressionist tradition which, in itself, forms part of the modernist movement.[6] His style can also be tentatively linked to the [French] impressionist style (e.g. Fig. 7.8). The link between artist and the movement is based on timing. The influence of the impressionist movement was still in vogue following World War I and chaos that ensued post-war and, over this time, would morph into artistic movements that kept a core ideology in what impressionism stood for and the protagonists who adopted and experimented with it.

Expressionism was initially a movement that involved the performance arts and originated in Germany at the beginning of the 20th century. It was a reaction to the dehumanisation caused from industrialisation and the growth of the *metropolis*. In political terms, it was a [Marxist] reaction from the repression of bourgeois materialistic values on the working classes (for want of a better term). The working class were considered to be the essential modes of production who were strictly controlled. It is from this standpoint that expressionists rejected the harsh realities of realism. It is probable that a history of bloodshed through nationalistic ideologies and armament on an industrial scale between 1905 and 1919 had entrenched a form of passivism,

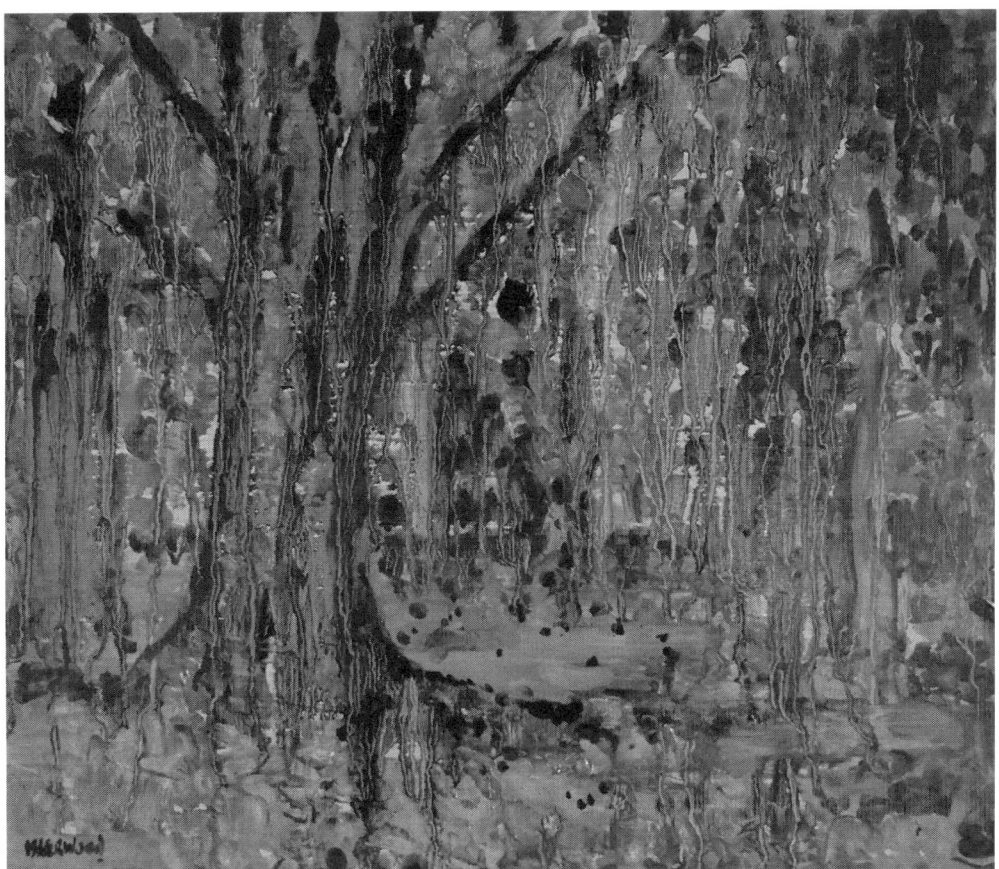

Figure 7.8: Isherwood's Wooded Glade *(Courtesy of the Pilkington Collection)*

even within those thinkers who had been directly involved in fighting during World War I. Even the haunting imagery by Surrealist painter Paul Nash rejected the realism of war by painting the aftermaths of the battlefield. His sharp shadows and splashes of paint representing the slaughter of the Western Front created a form of semiotic symbolism that exposed the horror of a total industrial war scenario; in other words, the symbol of realism without the portrayal of realism. Cynically though, one can argue that expressionism is present in every visual image in that the image represents *something*. Expressionists would argue that it is the degree of intensity that matters and how that intenseness arouses the onlooker; what the Italian writer and essayist Alberto Arbasino remarks 'expressionism throws some terrific *fuck yous*' (Pedullà 2003, 157)!

Although expressionism developed as a result of the avant-garde movement, the style accelerated during the fractious Weimar Republic in Germany, following German defeat in World War I. The centre for the expressionism movement was Berlin.[7] The expressionist style extended to a wide range of the arts, including architecture, dance, film, literature and music and theatre. In Aldous Huxley's *Brave New World* (1932) the sciences and, to some extent, the arts were moving to construct a Utopian world; however, the arts movement in reality was, during the latter part of the early 20th century, still experimental and somewhat naive. For example, it would take Isherwood until his 30s to understand his place in the art world. This above all led to his frustration and the coming-to-terms with social standing and angst within the sometimes malicious and conservative institution that was the *art world*.

A troubled style

Although the term 'expressionism' was first coined during the mid-19th century, its origins are firmly rooted in the critique for paintings that were exhibited in the Paris Salon in 1901 by the artist Julien-Auguste Hervé.[8] Hervé used the term *expressionismes* which was an alternative view to then popular style of *impressionism* which had taken the art world by storm 20 years previously. A decade after Hervé, artist Antonin Matějček claimed:

> An Expressionist wishes, above all, to express himself... (an Expressionist rejects) immediate perception and builds on more complex psychic structures... Impressions and mental images that pass through... people's soul as through a filter which rids them of all substantial accretions to produce their clear essence [...and] are assimilated and condense into more general forms, into types, which he transcribes through simple short-hand formulae and [semiotic] symbols. (cited in Gordon 1987, 175)

An important literary influence on expressionism was work by the German philosopher Friedrich Nietzsche where he describes the 'self' in his *Thus Spoke Zarathustra* (1883–92), Following Nietzsche's publication, a raft of intellectuals began to explore the philosophy of expressionism through drama and literature (e.g. August Strindberg, Frank Wedekind, Walt Whitman, and of course Sigmund Freud). In terms of art, early exponents of expression included Edvard Munch, James Ensor and Vincent van Gogh. It is more than likely that Isherwood and other British exponents to expressionism took their influences from these artists, usually from art college mentors who would have been at the forefront of artistic critique during the 20 years either side of 1900 when artistic expressionism was in its infancy.

Prior to World War I, the expressionist movement was gaining a reputation in Dresden, Germany, led by Ernst Kirchner who formed *Die Brücke* (translated as 'The Bridge'). This movement was later joined in 1910 by a group of Munich artists who formed *Der Blaue Reiter* (translated as 'The Blue Rider'). Although what we now know as [early] expressionism, the term was not officially used until 1913. During the inter-war years, expressionism had begun to wane in Germany due mainly to the rise in Fascism; however, it was taken up with vigour elsewhere in Europe and North America. It should be noted that at this time expressionism had been partly responsible for the birth of Cubism, Dadaism and Surrealism; what I would term as extreme artistry; here, the arts, along with new ways of thinking about the way we live were entering a *brave new world*!

The stratigraphy of a brave new world after total war for a second time

During the inter-war and post-war periods, expressionism in Britain was accepted, albeit on an *ad-hoc* basis, by a selected number of artists. I would argue, however, that many artists experimenting with artistic endeavour and philosophy during these politically turbulent times were already using expressionism in order to get their political point across through various artistic and literal mediums (e.g. The Mitford Sisters and the Bloomsbury Group). In terms of the visual arts, exponents such as Francis Bacon, Lucian Freud, Howard Hodgkin and John Walker and, of course, James Isherwood developed their own individually distinct styles through their chosen media. Despite the expressionist movement gaining pace at the turn of the 20th century, it was not until the post-war period that a British expressionism movement developed. This movement ran alongside a number of allied media including avant-garde Cubism and Surrealism. Post-war, one begins to witness the expressionist movement in Britain splintering into a numbers of related *isms*; L.S. Lowry and, later, Isherwood are influenced by *figurative* expressionism; a movement that was based on humanist philosophy and promoted both personal and group identity in the modern world. This branch was later adopted by [revivalist] Neo-expressionist movement of the latter part of the 20th century[9] and included protagonists such as Francis Bacon, Peter Blake and David Hockney[10] who incorporated people into scapes; the *self* and *persona* being the focus for the onlooker. This approach was, in view a reaction against post-war abstract works that focused on the *self* and the *body* but without context (e.g. Henry Moore and Jacob Epstein); roots that were firmly established in the Surrealism of the inter-war period. Concerning Isherwood's northern-style, the persona, the character is clearly fixed, usually as the primary focus within a recognised northern town scene. Indeed, I would argue that Isherwood is portraying himself in many of his paintings, reclaiming his upbringing and early working life. The backdrop – usually the northern town or factory scene – can be considered semiotic in that it promotes adjectives that relate to poverty and being working class.

Modernity and the retrospective

The expressionist movement and, in particular, exponents of the post-war British tradition (for want of a better phrase) comprised a series of narratives rather than a single movement or ideology. In many respects, I question the term 'expressionism'. By way of meaning, expressionism is confined to the *self* and what one expresses and how it is expressed. In the case of post-war critical thinking the movement can be considered a meta-narrative – a narrative embracing many narratives using a historic discourse and experience, resulting in an all-embracing concept (Carr 1986). Based on this rather loosely-defined definition, Isherwood can be considered to be part of this meta-narrative. If alive today, Isherwood might be inclined to agree. His northern-style reveals something of the past which is placed on canvas belonging to Isherwood's present; the meta-narrative here being one of nostalgia and humility with a degree of contempt to the class-ridden politics that enforced people like Isherwood's *Wigan Women* to reveal the poverty of a post-war generation; an issue that we can relate to with today's establishment populist politics.

Acknowledgement

The author would like to express sincere thanks to the Pilkington family for permission to take digital images from their James Isherwood Collection. All mistakes are of course my responsibility.

Notes

1. *Le Peintre de la Vie Moderne* (1863)
2. The provenance of this painting is disputed, the period in which was painted is not.
3. https://arquivo.pt/wayback/20090710115527/http://traumwerk.stanford.edu/philolog/2005/11/ekphrasis_ovid_in_pieter_breug.html
4. Date unknown.
5. For example, *Workers at Pilkington Glass Works* (c. 1900) and *Employees Leaving Vickers and Maxim's* (1901)
6. Embracing the early 20th-century Avant-Garde movement.
7. At the same time the Bauhaus movement in Germany was in its infancy.
8. One can argue that expressionism (by another name) has its roots within the European Renaissance, especially in the way landscape is portrayed (e.g. El Greco's *View of Toledo* (1592) and Pieter Bruegel's multi-narrative *Landscape with the Fall of Icarus* (1558[?]).
9. Otherwise known as the 'young contemporaries'.
10. Hockney was also considered by critics as an exponent of the British Pop-Art scene.

References

Carr, D. 1986. *Time, Narrative, and History*. Bloomington IN: Indiana University Press.

Foucault, M. 1975. *Surveiller et punir: naissance de la prison*. Paris: Gallimard.

Foucault, M. 1977. *Discipline and Punish: The Birth of the Prison*, translated by Alan Sheridan. London: Penguin Books.

Gordon, D.E. 1987. *Expressionism: Art and Ideas*. New Haven: Yale University Press.

Hodder, I. 2012. *Entangled: An Archaeology of the Relationships between Humans and Things*. London: Wiley Blackwell.

Huxley, A. 1932. *The Brave New World*. London: Chatto & Windus.

Iddon, B. 2013. *James Lawrence Isherwood (1917 - 1989)*. Cirencester: Memoirs.

Kilinski II, K. 2004. Bruegel on Icarus. Inversions of the Fall. *Zeitschrift für Kunstgeschichte* 67(1), 91–114.

Pedullà, G. 2003. Sull'albergo di ciliegie. Conversando di letteratura e cinema con Alberto Arbasino. *Contemporanea. Rivista di studi sulla letteratura e sulla comunicazione* 1, 147–160.

Chapter 8

The demography of prehistoric artists

Ezra Zubrow

People are often unaware of their own unawareness

(Gilovich *et al.* 2002)

This chapter is mostly a heuristic. Although it includes some real data, its greater value is in the questions raised in trying to write the chapter. It points out the stereotypes we create about artists and provides an interesting set of concerns about how society has changed.

The single overarching trend about prehistoric art is the degree of conservative consistency over time (Tedesco 2000). A brief review of images (Moser & Gamble 1998) and cartoons[1] of 'prehistoric artists' or the popular literature (Zimmer 2018) regarding 'prehistoric artists' demonstrates that they are the products of 'primitive modernity' (Pearson 2007) where modern characteristics are time-travelled into the past. They tend to be bearded, bereted, flaky, broke, wear dirty and ragged clothes, are perfectionists, jazz loving, eccentric, and moody while living in a dream state. Demographically they are expected to be male and young.

What are the questions that are raised when one begins to ask what are the demographic characteristics of prehistoric artists. Here are some of them:

- How do we define what is an artist?
- Should artists be limited to the visual arts?
- Should being an artist include music, dance, and other subjects?
- Does it need to be a full time occupation or can it be part time?
- Does it need to be professional (in the sense compensated) or can it be vocational? And what do these concepts mean for prehistoric societies?
- How does the definition of the artist's occupation change over the great biological span of human and hominid evolution?
- How does it change across the huge diversity of societies and cultural evolution?
- How does it change over the exceptionally varied geography of the earth?
- Do the numbers of artists increase or decrease across time and increase and decrease proportionally to the population?

- What are the changes in the numbers of artists, their ages, and their gender ratio relative to the society's population size as one moves from the Palaeolithic to the Mesolithic to the Neolithic and to historical and industrial societies? (we know that in modern industrial societies, as one gets older, the percentage of artists decreases as people decide that the employment or success opportunities are not sufficient and one meets past musicians and artists all the time).

Let us consider what an artist actually is. One discovers that it is a complex of occupations. In fact, there are over 800 occupations which could be included in the artist category. They can be categorised into some 23 major groups, 97 minor occupation groups, 461 broad occupations, and 840 detailed occupations.[2]

Examining these occupations[3] there are a variety of classificatory characteristics that need to be explored. Criteria for artists include the nature of the work performed, duties necessary, commonalities, distinctions from other occupations, educational requirements, credentials, technologies and professional associations, very broadly defined.

Does the definition of an artist require it to be a full time occupation? What does full time mean in various prehistoric situations? To be an artist, does the work performed need to be compensated by pay or profit? If so, how do we consider the prehistoric traditions when artistic occupations were not performed for monetary compensation or profit. However, how should we consider the long prehistoric traditions of receiving religious or social rewards? Does the artist need to do this full time and what would be full time during various prehistoric times? Some artists are volunteers. Should artists be considered in only one category? Many artists both today and prehistorically may fit into many categories simultaneously – a person may be a painter, a musician, a religious official and a businessman. For example, a Greek ceramicist paints bell-craters in the black figure style and trades them to a temple, in which he is a priest, for ritual use that includes musical rituals. Artists may be defined by the work performed, or by their skills and training, or by both. There may be a competence threshold that differentiates the artist from the 'dabbler' or amateur. Then, one should also consider the ancillary occupations and labourers. There are individuals whose primary occupations are planning, directing, or teaching artists. And similarly, there are apprentices, trainees, helpers and aides of a variety of sorts. Each might or might not be considered an artist.

Turning to general trends, in new demographic information such as age, education levels, income, ethnicity, and other social characteristics; like most people I had some general stereotypes. My expectation prior to this research was that the total number of artists is probably increasing simply because the population has increased so much over time (Robinson 2003; Figs 8.1 & 8.2). This is the case, for example, in the US where the numbers of artists are growing (Fig. 8.2). The exact numbers may vary depending upon the analyst but for our purposes that is irrelevant.

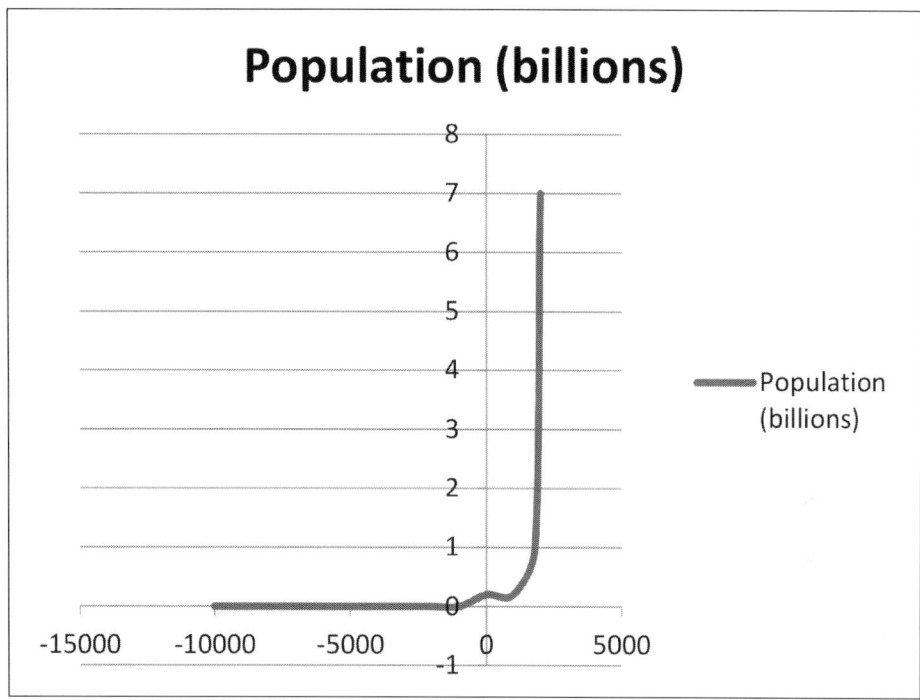

Figure 8.1: The growth of world population over the last 12,000 years

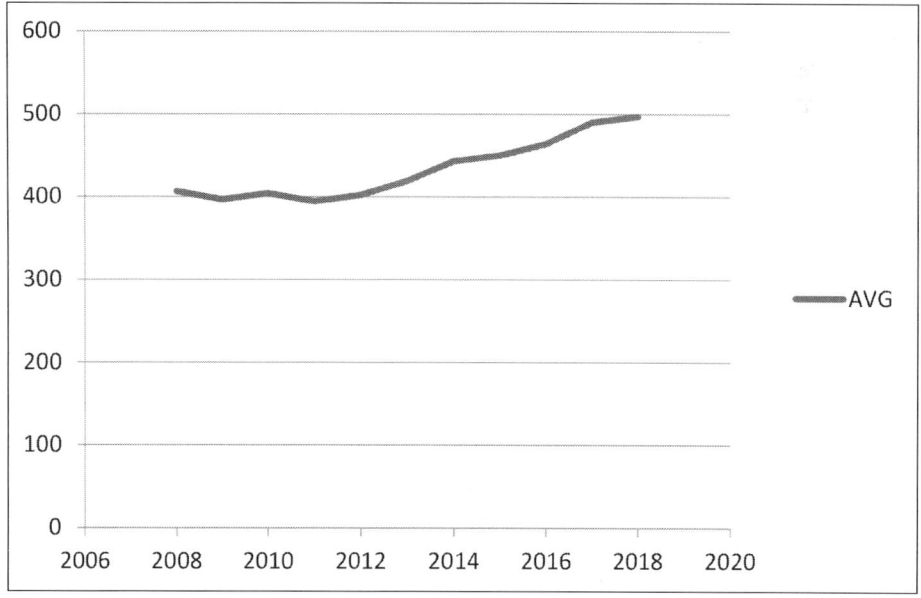

Figure 8.2: The average yearly number of artists in the United States by year (based on averaging monthly counts)

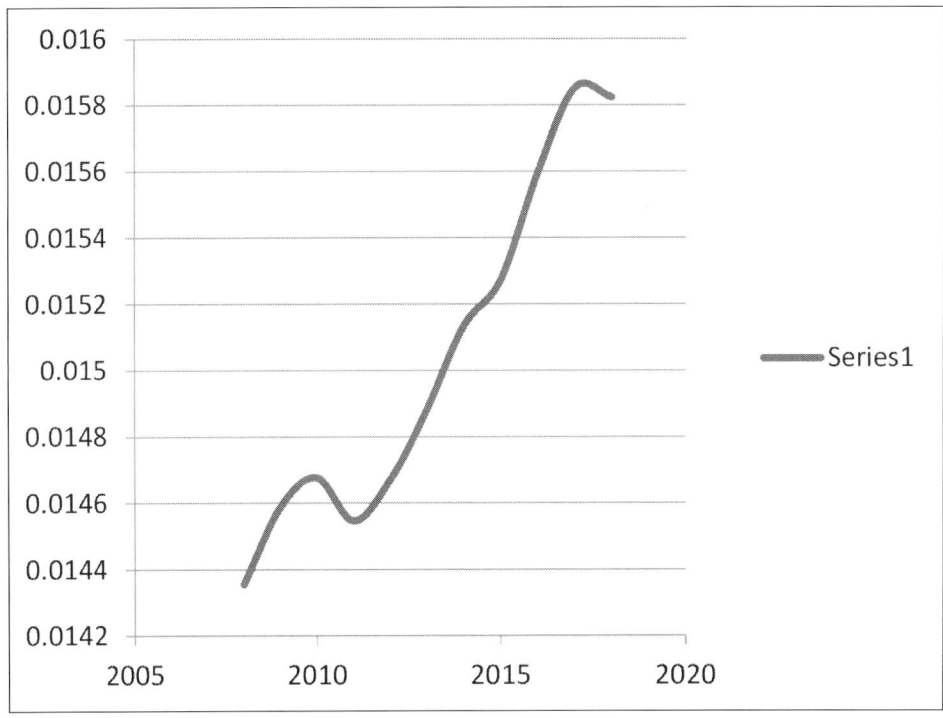

Fig. 8.3: The annual average percentage of US employees in arts, entertainment and recreation (not seasonally adjusted) as a percentage of all US employees (not seasonally adjusted) (note Nov. and Dec. 2018 were preliminary values)

I also believed that, relative to the population as a whole, the proportion of artists is decreasing. I assumed this because I thought the importance of the arts is diminishing in our increasingly technological, industrial and entrepreneurial societies. I was surprised to find that it was not the case. In Figure 8.3, one sees that even proportionally the number of artists is increasing.

Now, turning from the numerically accurate to the inaccurate, I will ask the reader to temporarily suspend disbelief. In the following there is much to criticise in terms of assumptions, in expression of analogies and in the choice of proxies. One cannot assume that the following results are anything but a heuristic to indicate interesting trends that may have some reality. However, they would need far more data and investigation to be in any way conclusive.

How do we, methodologically, find the numbers, ages and genders of prehistoric artists? It is difficult. Having searched the literature there are almost no articles providing data. One might consider using a proxy similar to that which has been done in general prehistoric demography. Frequently, one uses such data as the number of sites, of structures, of fire pits, the number or area of houses. Other proxies have been the division of the size of cooking vessels by the size of the consuming dishes.

One could try, analogously, to use the number of artistic artefacts produced at a site or within a culture as a proxy. For example, we know that the recent[4] artist Van Gogh produced more than 2000 artworks, consisting of around 900 paintings and 1100 drawings and sketches – averaging more than 54 a year for his entire 37 years. Rodin is thought to have created around 17,000 pieces that are in various places around the world. This would average some 220 a year for his 77 years. Similar figures for Botticelli are more than three a year and we know that he produced more than nine complete pictures in 1470 alone. One might be able to estimate an approximate production range and work backwards from the number of artistic pieces to the number of artists. There are over 15,000 individual pieces of artwork in Tassili n'Ajjer National Park, Algeria, but they are not contemporaneous. The art includes paintings and engravings into the rock that depict the culture of Africans up to 12,000 years ago. Alternatively, count the number of *ateliers* (workshops or studios). In Alta, Norway, there are some 6000 prehistoric rock carvings and paintings. They are located at five separate sites along the Altafjord.[5] Similarly, looking at sites, one might note that there are references to Ptolemaic glass works in the Alexandria region but only four workshops have been identified that specialised in making both monochrome and millefiori inlays (Boschetti 2018). One begins to see the scope of the methodological problem.

The *Human Relation Area Files* (eHRAF) are a database of some 321 ethnographic cultures begun during World War II. Its purpose is 'to support and conduct original

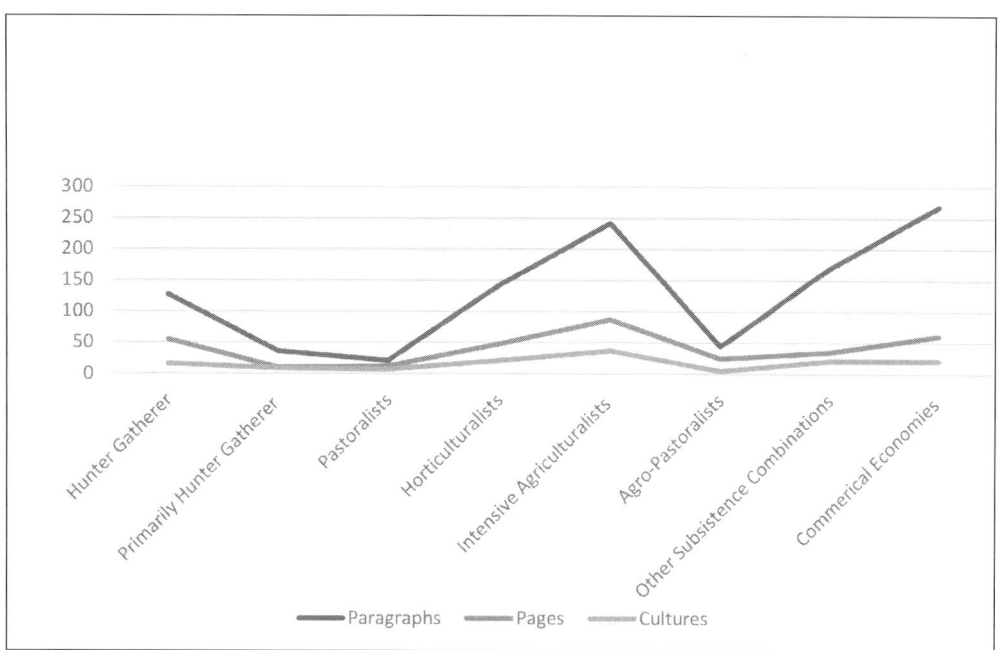

Figure 8.4: Proxy for the importance and number of artists by subsistence type

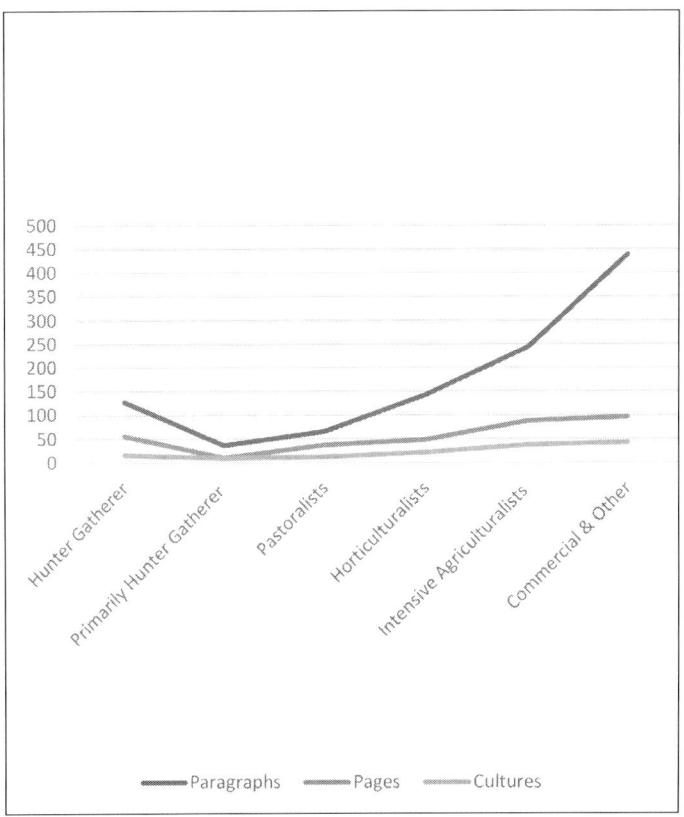

Figure 8.5: Proxy for the importance and number of artists by subsistence type combining Pastoralist categories and then Commercial and Other

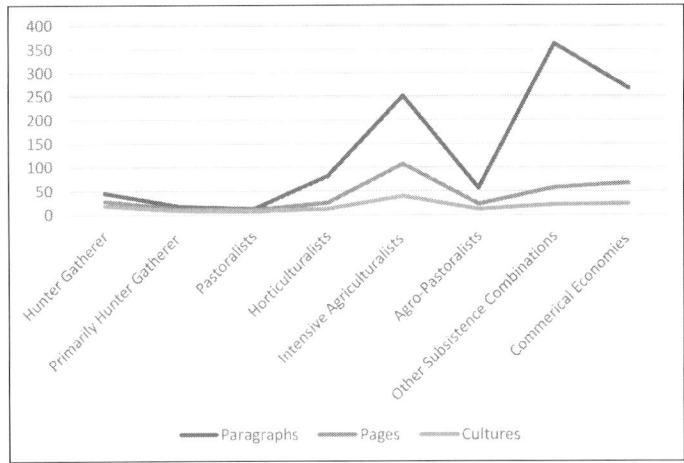

Figure 8.6: Proxy for the importance and number of musicians by subsistence type

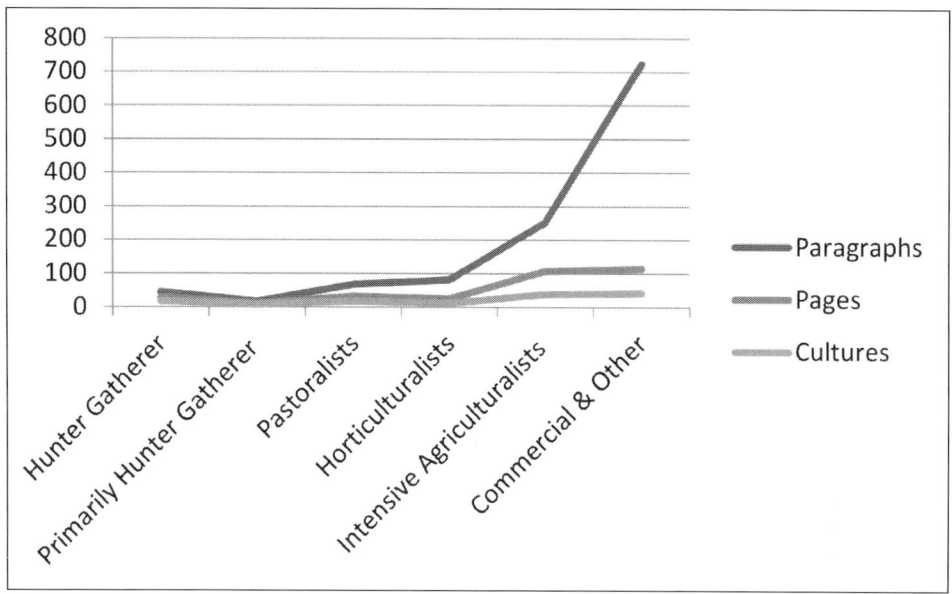

Figure 8.7: Proxy for the importance and number of musicians by subsistence type combining Pastoralist categories and then Commercial and Other

research on cross-cultural variation'. I will use it as a very rough proxy of the demo-graphic importance of artists in different types of societies. I also will make a very rough ethnographic analogy between archaeology and types of society.

Let us assume that there is a general chronological order to world prehistory. Namely, Palaeolithic precedes Mesolithic, Mesolithic precedes Neolithic, Neolithic precedes the Metal Ages, and the Metal Ages precede the Industrial Revolution. Similarly, let us assume that Hunter Gatherers precede Primarily Hunter Gatherers, Primarily Hunter Gathers precede Pastoralists, Pastoralists precede Horticulturalists, Horticulturalists precede Intensive Agriculturalists, Intensive Agriculturalists precede Agro-Pastoralists, Agro-Pastoralists precede Other Subsistence Combinations, and Other Subsistence Combinations precede Commercial Economies. Let us suspend dis-belief once more and accept an assumption that the importance of a cultural feature, such as artists, is reflected in the ethnographies. If we accept this we may construct Figure 8.4, showing a general upward trend that is punctuated by diminished numbers as one moves into the pastoralist categories.

One sees a general upward trend that is punctuated by diminished numbers as one moves into the pastoralist categories. This is clearer if one combines the Pastoralist categories (Fig. 8.5). Looking at a single category of artists-musicians one sees the pattern shown in Figure 8.6. Similarly, recombining the number of Pastoralists and then Commercial Economies with Other Subsistence patterns as we did with artists, the increasing trend post Pastoralism is very clear (Fig. 8.7). The geographic distribution

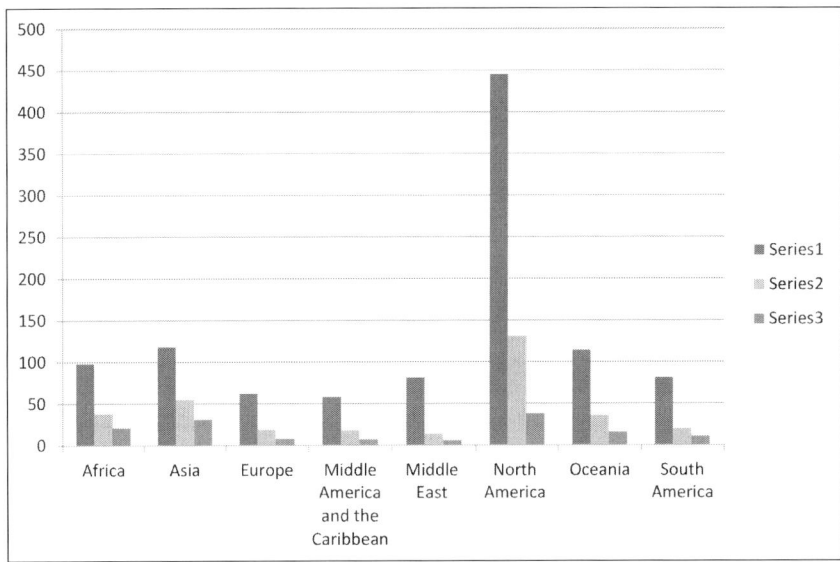

Figure 8.8: Proxies for the number of artists by continent among anthropological societies based upon eHRAF

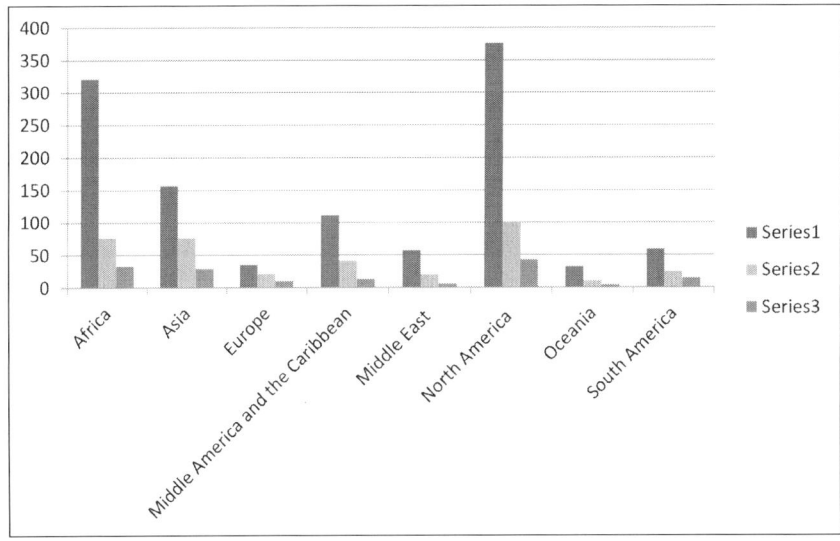

Figure 8.9: Proxies for the number of musicians by continent among anthropological societies based upon eHRAF

of artists across all the subsistence types is shown in Figure 8.8 and the geographic distribution of musicians is illustrated in Figure 8.9.

Three clear and interesting inferences may be made from these distribution illustrations. One is the dominance of North America in both Artists and Musicians. The second is how much more dominant Africa is in the musicians than in the artists. Finally, one notes the relatively small proportion that Europe represents in contrast to its reputation as historically the centre of artistic culture.

Conclusion

I found asking the question 'what are the definitions and characteristics of the prehistoric demography of artists' more valuable than the actual results. It is a very useful heuristic approach.

- First, this chapter points out the 'conservative consistency' in the modern popular view of 'prehistoric art and prehistoric artists' and its attribution of 'modernity' to prehistory.
- Second, it shows how complicated is the question of 'what is an artist'. Although it is a very diverse profession, the range of different occupations may be specified. One is able to get an insight into the diversity and possible combinations. This results in a far better understanding of the types of artists one could be concerned about at different periods of time and in different places.
- Third, my initial stereotypes – that the number of artists is increasing over time and decreasing relative to the population as a whole – is not the case in at least one large modern nation for which there is accurate data; the number of artists is increasing and is increasing relative to the entire population.
- Fourth, there is a discussion of the use of different methodologies and proxies for the number of prehistoric artists. One suggestion is that, analogously to the use of the number of prehistoric rooms and prehistoric sites to estimate prehistoric population, one might use the number of prehistoric artistic artefacts and the number of prehistoric workshops.
- Fifth, using ethnographic analogy one is able to show a generally increasing number of prehistoric artists across time from the hunting and gathering Palaeolithic through the Mesolithic, the Neolithic and the various metal ages into the modern commercial and industrial economies, with the notable exception of the periods of pastoralism. This is shown for both artists and musicians.
- Sixth, geographically our data shows a dominance of North American artists and musicians. However, in the case of musicians, the numbers between Africa and North America are far closer and, in both artists and musicians, the numbers from Europe are remarkably small.
- Finally, one needs to re-emphasise that most of the data in this chapter are proxies and should not be used in any way as conclusive. Instead, the value of this chapter is in the questions it raises and the indicators that one might use for future research. As the famous Hungarian philosopher of mathematics and science, Imre Lakatos said 'Research programs, besides their negative heuristic, are also characterized by their positive heuristic' (Musgrave & Pigden 2016). I hope that this chapter is part of that affirmative heuristic.

Notes

1 Google search 'Cartoons prehistoric artists' 14 May 2019 https://www.google.com/search?q=pre-historic+artists+cartoons&tbm=isch&source=univ&sa=X&ved=2ahUKEwi-u-SZ-ZriAhV-CAqwKHSEQBJUQsAR6BAgIEAE&biw=1344&bih=639
2 Based on the 2010 *Standard Occupational Classification of the US Bureau of Labor Statistics* https://www.bls.gov/soc/2010/#classification. There is a 2018 Standard Occupational Classification but it is generally similar for artists. https://www.bls.gov/soc/2018/major_groups.htm
3 1. Artists and related workers
    1.1. Art directors
    1.2. Craft artists
    1.3. Fine artists, including painters, sculptors, and illustrators
    1.4. Special effects artists and animators
    1.5. Artists and related workers, all other
  2. Designers
    2.1. Commercial and industrial designers
    2.2. Fashion designers
    2.3. Floral designers
    2.4. Graphic designers
    2.5. Interior designers
    2.6. Merchandise displayers and window trimmers
    2.7. Set and exhibit designers
    2.8. Designers, all other
  3. Entertainers and performers
    3.1. Actors
    3.2. Producers and directors
    3.3. Dancers
    3.4. Choreographers
    3.5. Music directors
    3.6. Composers
    3.7. Musicians
    3.8. Singers
  4. Writers and editors
    4.1. Editors
    4.2. Technical writers
    4.3. Writers
    4.4. Authors
  5. Photographers
4 It gives me considerable pleasure and I can think of very few other contexts in which one would describe Van Gogh as recent.
5 I had a chance to help excavate a few of these. As one pulled back the moss and the art appeared as clear and fresh as when it was made, there was a sense of discovery and awe that was second to none.

References

Boschetti, C. 2018. Working glass in Ptolemaic Egypt, a new evidence from Denderah. *Journal of Archaeological Science Reports* 22, 550–558.
Gilovich, T., Griffin, D., & Kahneman, D. (eds) 2002. *Heuristics and Biases: The psychology of intuitive judgment.* Cambridge: Cambridge University Press.

Moser, S., & Gamble, C. 1998. *Ancestral Images: The iconography of human origins.* Ithaca, New York: Cornell University Press. Retrieved from http://www.jstor.org/stable/10.7591/j.ctv5rf35f.14

Musgrave, A. & Pigden, C. 2016. Imre Lakatos. In E.N. Zalta (ed.), *The Stanford Encyclopedia of Philosophy* (Winter 2016 edn). [https://plato.stanford.edu/archives/win2016/entries/lakatos/]

Pearson, R. 2007. Primitive modernity: H.G. Wells and the prehistoric man of the 1890s. *Yearbook of English Studies* 37(1), 58–74. [JSTOR, www.jstor.org/stable/20479278]

Robinson, W.C. 2003. Demographic history and theory as guides to the future of world population growth. *Genus* 59(3/4), 11–41. [JSTOR, www.jstor.org/stable/29788773]

Tedesco, L.A. 2000. Introduction to prehistoric art, 20,000–8000 BC. In *Heilbrunn Timeline of Art History*. New York: Metropolitan Museum of Art. [http://www.metmuseum.org/toah/hd/preh/hd_preh.htm (August 2007)]

Zimmer, C. 2018. The Neanderthal, the artist: Recent studies of cave art suggest that the cousins of modern humans were more sophisticated than their beastly reputation. *New York Times* 167 (57886), D1–D6.